PRINCIPAL MONUMENTS, BUILDINGS, GARDENS, ESTATES AND COASTAL AREAS

Clumber Park

LINCOLNSHIRE

Gunby Hall

ardwick Hall

DERBY-SHIRE

NOTTINGHAM-SHIRE

Tattershall Castle

Kedleston Hall

Sudbury Hall

Belton House

Calke Abbey

Felbrigg Hall

Blicklin

Oxb

LEICESTERSHIRE

Peckover House

ST LANDS

WARWICK

Packwood House

Baddesley Clinton

oughton Park

Charlecote House

ote or den

Farnborough Hall

Canons

Ashby House

Upton House

Stowe Landscape Gardens

Claydon House

Ascott

vshill Chastleton or

erborne k

OXFORD-SHIRE

Waddesdon Manor

dworth an

Buscot Park

Ashdown House

Grey's Court

vebury

Basildon Park

NORTHAMPTON-SHIRE

CAMBRIDGE-SHIRE

Anglesey Abbey

Wimpole Hall

BEDFORD-SHIRE

HERTFORD-SHIRE

Ashridge Estate

BUCKS

West Wycombe Park

Hughenden Manor

Cliveden

Fenton House

Osterley Park

LONDON

Ham House

SUFFOLK

Ickworth

Melford Hall

Flatford Mill

Paycocke's

ESSEX

Claremont Landscape Garden

BERKSHIRE

Sutton House

KENT

LTSHIRE

The Vyne

HAMPSHIRE

mpesson se

Hinton Ampner

Mottisfont Abbey

SURREY

Hatchlands Park

Clandon Park

Winkworth Arboretum

Polesden Lacey

Chartwell

Standen

Nymans

Petworth House

Uppark

Knole

Ightham Mote

Quebec House

Scotney Castle

Sissinghurst Castle

Smallhythe Place

Bateman's

Sheffield Park

Bodiam Castle

Lamb House

gston Lacy

WEST SUSSEX

EAST SUSSEX

Alfriston Clergy House

FROM ACORN TO OAK TREE

JENNIFER JENKINS AND PATRICK JAMES

FROM ACORN TO OAK TREE

THE GROWTH OF THE NATIONAL TRUST
1895–1994

M

MACMILLAN

LONDON

First published 1994 by Macmillan London

a division of Pan Macmillan Publishers Limited
Cavaye Place London SW10 9PG
and Basingstoke

Associated companies throughout the world

ISBN 0 333 58953 X

Copyright © Jennifer Jenkins and Patrick James 1994

The right of Jennifer Jenkins and Patrick James to be identified as the
author of this work has been asserted by them in accordance
with the Copyright, Designs and Patents Act 1988.

1 3 5 7 9 8 6 4 2

A CIP catalogue record for this book is available from
the British Library

Photoset by Parker Typesetting Service, Leicester
Printed and bound in Great Britain by Mackays of Chatham PLC, Kent

CONTENTS

LIST OF ILLUSTRATIONS   vii
ACKNOWLEDGMENTS   ix
PREFACE BY JENNIFER JENKINS   xi

1   The Founders   1
2   Early Years   24
3   Urban Sprawl   43
4   The Country-House Scheme   74
5   Subtopia and the Second World War   104
6   Peace and Socialist Utopia   122
7   White Elephants   145
8   Western Approaches   165
9   Gardens and Quirkeries   188
10   Admiral Neptune and Commander Rawnsley   219
11   Trial by Benson   240
12   The New Professionals   258
13   Decorating Historic Houses   269
14   Fora, Fauna and Flora   285
15   A Personal Perspective by Jennifer Jenkins   307

EPILOGUE   331

APPENDIX A
Officers of the Trust, 1895–1994
335

APPENDIX B
Trust Membership, Acreage and Staff, 1895–1990
337

REFERENCES   338

INDEX   351

LIST OF ILLUSTRATIONS

Between pages 114/115

1. Octavia Hill, 1882 (National Trust Photographic Library)
2. Sir Robert Hunter, c.1890 (Guildford Muniment Room, Surrey Record Office)
3. Hardwicke Rawnsley with his son Noel and Beatrix Potter, 1885 (Rosalind Rawnsley)
4. Duke of Westminster by Ape, 1870 (Hulton Deutsch Picture Library)
5. John Bailey, 1926 (National Portrait Gallery)
6. G. M. Trevelyan and Sir Charles Trevelyan, 1945 (National Trust)
7. Vita Sackville-West and Harold Nicolson, 1938 (Nigel Nicolson)
8. Evelyn, Duchess of Devonshire, 1950 (Chatsworth Settlement Trustees)
9. Earl of Crawford, Jack Rathbone, Earl De La Warr and Hubert Smith at an Annual General Meeting in the 1950s (National Trust)
10. Oliver, Viscount Esher with his wife, 1954 (The present Viscount Esher)
11. James Lees-Milne, 1975 (James Lees-Milne)
12. Queen Elizabeth, the Queen Mother, and the Earl of Antrim, 1954 (The Hon. Randal McDonnell)
13. Cubby Acland and tenant farmer, 1977 (National Trust, North West Region)
14. Jack Boles (National Trust)
15. Lord Gibson (National Trust)
16. Angus Stirling, 1993 (Richard Holttum)
17. Philip Yorke (National Trust)
18. Henry Harpur-Crewe, 1984 (National Trust)
19. Clergy House, Alfriston, East Sussex, 1894 (National Trust)

Between pages 242/243

20. Skiddaw reflected in Derwentwater, Lake District (National Trust)
21. View of Winnats Pass from Mam Tor, Peak District (National Trust)
22. Wicken Fen, Cambridgeshire (National Trust)

23. Luccombe, Holnicote estate, Somerset (National Trust)
24. Stourhead, Wiltshire (National Trust)
25. Hidcote Manor Garden, Gloucestershire (Graham Stuart Thomas)
26. Blickling Hall, Norfolk (National Trust)
27. Sudbury Hall, Derbyshire (Country Life Picture Library)
28. Cragside, Northumberland (National Trust)
29. Cliffs at Kynance Cove, Cornwall (National Trust)
30. View from Plas Newydd, Anglesey, to Snowdonia (National Trust)
31. Stone Circle, Avebury, Wiltshire (National Trust)
32. Fountains Abbey from Studley Royal Garden, North Yorkshire (National Trust)
33. Quarry Bank Cotton Mill, Styal, Cheshire (Mike Williams)
34. Souter Lighthouse, South Shields, Tyne and Wear (National Trust)
35. Uppark, West Sussex (National Trust)
36. Kingston Lacy, Dorset (National Trust)

ACKNOWLEDGMENTS

THIS BOOK could not have been written without the help of many past and present members of the National Trust's staff and committees and others connected with the Trust. We thank all these and in particular the following who have commented on one or more sections of the book: the Marquess of Anglesey, Nicholas Baring, Lord Benson of Drovers, Sir Frederick Bishop, Ivor Blomfield, Sir Jack Boles, Sir Charles Brett, Lord Chorley, Len Clark, John Cornforth, Spencer Crookenden, Martin Drury, Edward Fawcett, Lord Gibson, John Harvey, Robert Latham, James Lees-Milne, Stanley Morris, Mark Norman, Jacob Simon, Sir Angus Stirling, Graham Stuart Thomas, David Thackray, Michael Trinick and Sir Marcus Worsley.

We are greatly indebted to Janette Harley, the National Trust's archivist, for her skill in tracking down documents and files and to her patient assistants, Jill Casson, Mark Perry, Andrew Procter and Victoria Thomas; to Samantha Wyndham and Gayle Moult for help with photographs; and to Iain Wilson for the compilation of statistics.

We are grateful to the Earl of Crawford and Balcarres, the Earl of Rosse, the Marquess of Zetland and the Hon. Randal McDonnell for permission to quote from their family papers, and to the custodians of these papers at the National Library of Scotland, the Birr Scientific and Heritage Foundation, the North Yorkshire County Record Office and the Northern Ireland Public Record Office respectively.

Thanks are due for permission to quote from *Octavia Hill: A Life* by Gillian Darley (Constable); *Figures in a Landscape* by John Gaze (Barrie & Jenkins); *Ancestral Voices, Prophesying Peace* and *Caves of Ice* by James Lees-Milne (Chatto & Windus); *People and*

Places by James Lees-Milne (John Murray); *Brideshead Revisited* by Evelyn Waugh (Peters, Fraser & Dunlop).

We also wish to thank many people who have let us have recollections and documents, including Viscount Esher, Lord Beaumont of Whitley, Len Clark, Frances Fedden, John Grigg, Trevor Hussey, Commander Conrad Rawnsley, the Hon. Miriam Rothschild, Sir John Smith, Audrey Urry and the Ernest Cook Trust.

We have received invaluable assistance from Douglas Dow, George Russell, Sir James Stormonth Darling and the Earl of Wemyss and March of the National Trust for Scotland; and from Damaris Horan and Nathan Hale of the Royal Oak Foundation in the United States.

We are grateful to Alexandra O'Bryan-Tear and Deirdre Chappell for typing the manuscript, to Richard Burnett for research and to Alan Gordon Walker, Georgina Morley and Tanya Stobbs for their support at Macmillan. Peter James has exercised his unerring eye on the editing and Douglas Matthews has compiled the index with his unrivalled skill.

Finally, this book would not have existed but for Roland Philipps, who suggested that we write it and whose encouragement ensured its completion.

PREFACE BY JENNIFER JENKINS

As I came to know the National Trust I became increasingly fascinated by the question of how this private charity, wholly independent of government, had in the course of a hundred years become Britain's largest private landowner. Even more interesting is the question why the Trust should own landscape and historic buildings on a scale which in other countries would be undertaken only by the State.

If you visit the Grand Canyon or Yellowstone in the United States you will be on land owned by the federal government. But if you walk along the coastal path of Cornwall or the shores of Derwentwater, and if you climb the peaks of North Wales or study the wildlife of Strangford Lough in Northern Ireland, you are likely to find yourself on land owned by the National Trust. In France if you visit historic sites and buildings you will probably find that they are owned by the government or the municipality. But if you wish to explore British history through its buildings you can study properties of every period which are owned by the Trust, starting with the pre-historic stone circle at Avebury, continuing with the Roman Wall in North-umberland and the monastic ruins of Fountains Abbey, and then touring country houses with their collections of furniture and objects of art accumulated over six centuries.

In tracing the origins and history of the National Trust I have tried to discover some of the reasons why conservation in the United Kingdom has developed on lines distinct from those elsewhere. Fundamentally I think that it is the cultural attitudes and institutional traditions, evolved since the seven-teenth century, which have caused the British to seek solutions to environmental problems, as earlier to social problems, through voluntary associations rather than through state ownership. In addition, the social and economic changes of the past century have contributed to the growth of the Trust, particularly the decline – accentuated by two world wars – of the landed classes. This has been accompanied by the increasing affluence of much of the rest of the population, expressed in widespread car ownership and growing leisure.

Extensive though the Trust's estate has become it should not be exaggerated. Although it owns nearly 600,000 acres, 550 miles of coastline, and nearly 400 Sites of Special Scientific Interest, this amounts to little more than 1 per cent of England, Wales and Northern Ireland. In the United States national parks and other protected areas account for some 7 per cent of a much larger territory. In France public authorities own 8800 of the most important monuments and historic buildings – almost two-thirds – compared with only 300 or so owned by the Trust. Although the Trust also owns about 6 per cent of all archaeological sites and a large number of vernacular buildings, the Church of England is the largest owner of medieval buildings, and most country houses, including some of the grandest, are still in private hands. Despite the lack in Britain of the level of public intervention exercised in France and other continental countries, government agencies own the most visited national monuments – the royal palaces, Stonehenge and most of the Welsh castles – while local authorities have been responsible for most urban conservation. Other voluntary organizations also play an important part. The Royal Society for the Protection of Birds owns extensive bird sanctuaries, the preservation trusts have restored hundreds of historic buildings, the Council for the Protection of Rural England has campaigned tirelessly for the countryside and the amenity societies have lobbied effectively for the architectural heritage; all are supported by countless local societies.

But the National Trust is the only organization, official or voluntary, to span the whole range of 'places of historic interest or natural beauty' in England, Wales and Northern

Ireland. In this brief history, whose research and writing Patrick James and I have shared, I have tried to explain how it has reached this position and what have been the contributory economic and political factors. I have for convenience referred to it as the English Trust; the existence of a separate Scottish Trust renders the word British inappropriate.

JENNIFER JENKINS

CHAPTER ONE

The Founders

O N 12 JANUARY 1895, the National Trust for Places of Historic Interest or Natural Beauty was formally constituted. Conceived in 1884, it had been the vision of three pioneers – Octavia Hill, the housing reformer, Sir Robert Hunter, solicitor to the Commons Preservation Society, and Canon Hardwicke Rawnsley, a Lake District clergyman – who had become dedicated to saving land and buildings for the nation. In 1895, with the Duke of Westminster as president, they achieved their first objective: the incorporation of the Trust. Thus from the start were involved the complementary though not always coincidental interests of radical, reforming nature lovers and public-spirited landed grandees which has been one of the continuing features of the Trust.

In 1895, Britain was approaching the zenith of empire. In the summer of that year Lord Salisbury's third government began a decade of Unionist power, which enjoyed the united support of the landed interest in a way that would not have been possible before Gladstone's conversion to Home Rule had driven out of the Liberal Party the likes of Trollope's Duke of Omnium, who in the 1870s had regarded Toryism as socially rather inferior. Although the landed interest was united, its political pre-eminence had already been undermined by the Reform Acts. And in 1879 its mid-nineteenth-century prosperity, which had been enhanced by gains from railway development, mineral rights and the rising value of urban property, had come to an end with the onset of the long agricultural depression following the import of grain and meat from the New World. Many landed estates had seen their rent rolls fall by a third or more, and then in 1894 the introduction of death duties struck a further blow. The owners of great houses who found themselves without sustenance from business or American wives were increasingly forced into selling land or works of art.

Britain's industrial leadership was already under challenge. Producing more than half the world's steel in 1870, the country had been overtaken in this respect by both Germany and America. The 'deep red glow of [the] furnace fires'[1] of industrial England, glimpsed by Mr Pickwick on a visit to Birmingham in the 1820s, had begun to fade, but the growth of industry had brought about a massive movement of population from the countryside to the towns. By the end of the century 77 per cent of the population of England lived in towns and cities, compared with only 25 per cent a hundred years earlier.

The founders of the Trust had already shown their concern for the conditions in which the new urban dwellers lived and had been active in seeking to preserve open spaces to which they might escape. Robert Hunter, Octavia Hill and Hardwicke Rawnsley each contributed part of the essential vision and they received somewhat different but nonetheless invaluable support from the Duke of Westminster. As the early Trust bore contributory footprints from all four of them it is desirable to set out something of their varying backgrounds, interests and achievements.

Hunter was born in 1844 on Denmark Hill, London, the son of a prosperous merchant. A delicate child, he seemed unlikely to survive infancy. By 1860, however, he was strong enough to be taken with his sister to the Highlands. It was here that he acquired an interest in botany which was to develop into a passionate love of nature. At seventeen he went to University College, London, and by the age of twenty-two he had gained first-class degrees in both logic and moral philosophy. He was

next articled to a firm of Holborn solicitors. His career was to
change course when his uncle drew his attention to a competi-
tion offering several prizes amounting to £400 (enormous for
the time, the equivalent of about £14,000 today) for the best
essays on the preservation of commons. 'Here, Bob, you're
fond of scribbling, you'd better go in for this.'[2] The competi-
tion had been devised by Sir Henry Peek, who at the time was
active in defending Wimbledon Common, and among the
judges was George Shaw-Lefevre, the founder of the Commons Preservation Society(CPS). Although not the winning
entrant, Hunter's 15,000-word essay was one of those selected
for publication in *Six Essays on Commons Preservation*. He had
revealed a mastery of the complicated workings of commons
and, much to Shaw-Lefevre's satisfaction, had taken up the
CPS's line in contesting the rights of landowners. In his essay
he wrote:

> The mode of procedure will of course differ in different cases,
> but one suggestion may be made. Any Commoner whose
> rights are molested is clearly entitled to throw down the
> whole fencing or other obstruction erected. It may also be
> mentioned that in many cases, the Crown has rights of forest
> over waste lands, as, for example, over Epping Forest, and if
> these rights were properly insisted on, approvements (i.e.
> enclosures) must in such places cease. Every possible step
> should at once be taken; and the owners of the soil, if they
> persist in acts of violence, should be forced before a Court of
> Law or Equity.[3]

In 1867 Hunter joined the law practice of Philip Laurence,
whom, in the following year, at the age of only twenty-four, he
succeeded as senior partner and as honorary solicitor to the
CPS. It was a position for which he was ideally suited. The
success of legal actions taken to defend commons and open
spaces depended on proving the existence of the rights of
commoners. This involved collecting, sifting and marshalling a

3

mass of records and assembling witnesses to attend the trial – no less than eighty-three in the case of Epping Forest. It was Hunter's 'legal acumen, intense love of beauty, staunch loyalty to ideals once laid down and accepted, and a limitless patience with the paltry steps necessary to attain them'[4] that helped to preserve many of London's green spaces as well as a score of commons.

A staunch Liberal and a supporter of women's suffrage, he could express his idealism in his work for the open-space movement without placing himself directly in the heat of politics. His creative use of the law is described by his daughter thus:

> A less typical lawyer than Robert Hunter it would be hard to find. No layman could be more calmly scornful of the red tape in which the profession so often delights. His guiding principle as a lawyer was that the law is intended, and must be made, to serve the public good: and that when public interests and legal technicalities appear to clash, it is the duty to harmonize them if possible, or where they are incompatible, to get the law amended at the earliest possible date.[5]

His first wife died in 1870, and eight years later he married Ellen Cann, by whom he had three daughters; the youngest of these, Dorothy, remained an authoritative member of the Trust's Executive Committee and Council until the mid-1960s. She remembered him as sociable, fond of music and travel, and a strong churchman, with a great dislike of dogmatism and an unshakeable faith in the power of goodness. A few years after this second marriage he bought Meadfields, a large house in Haslemere, Surrey, where he lived for the next thirty-two years and from where he later instigated the National Trust's acquisition of the Devil's Punch Bowl, Box Hill and several other much frequented open spaces.

In 1882 Hunter relinquished his position at the CPS and

was appointed solicitor to the Post Office by Henry Fawcett, the blind Postmaster-General in Gladstone's second government and a vice-president of the CPS. His work there, although not so much in the public eye, was equally impressive and was recognized by a knighthood in 1894. He was responsible for more than fifty general Acts of Parliament, and through his negotiation of the Conveyance of Mails Act and his improvements to the efficiency of the Post Office Savings Bank he saved the public purse an estimated £10 million. *The Times* noted that 'he made it his rule to form his own conclusions and to give his own advice instead of opinion obtained from outside'.[6]

From 1875 Octavia Hill also took up the fight to preserve open spaces and it was in the offices of the CPS that she first met Robert Hunter. Born in Wisbech in 1838, she came of a family already noted for its public service. Her maternal grandfather, Dr Thomas Southwood Smith, had campaigned vigorously for public health reform and sanitary legislation. Her father, James Hill, was a corn merchant and banker who had founded a local radical newspaper. Octavia's mother Caroline was a teacher and a writer on educational theory. Octavia was her father's eighth daughter, although only her mother's third, and from an early age she possessed an 'absolute sureness of her own convictions . . . her dominant personality oddly at variance with her shyness and uncertainty'.[7] Her birth coincided with a downturn in her father's fortunes and by 1841 he was bankrupt. Two years later he suffered a nervous collapse and was advised by his doctor to separate from his family.

Caroline Hill moved her family to Finchley, then a rural village on the northern edge of London, with financial support from her father. There she was able to try out her educational theories on her daughters, giving them the benefits of open air, encouraging them to read widely and leaving them free to develop their own talents. This idyllic life came to an abrupt end when Octavia was thirteen and her mother was appointed to manage the Ladies' Guild, a crafts workshop for unskilled women and girls in Marylebone. From that time Octavia contributed to the somewhat precarious family income, first as manager of a group of young toymakers and then as secretary of the women's classes, which she attended at the Working Men's College. This had been founded by F. D. Maurice, the Christian Socialist, and attracted a number of eminent lecturers including John Ruskin.

At the Ladies' Guild Octavia met Ruskin, who was then the acknowledged arbiter of the nation's taste, and he was so impressed with her talent that he offered her paid employment copying illuminations and Old Master paintings. This exacting apprenticeship, enlivened by the conversation of her remarkable mentor, lasted ten years and led to a relationship which she described as being 'of a special order'.

It was Ruskin who in 1865 enabled her to embark on her life's work by lending her the money to buy two small blocks of filthy and dilapidated slum properties. She had become preoccupied with the degrading and unhealthy conditions in which the poor of Marylebone lived and had observed the improvements effected by a local charitable trust. Her aim was to put her properties into decent condition and to manage them on a sound financial basis. The weekly rent collections were occasions not merely to insist on prompt payment but to forge personal relations with the tenants. 'My strongest efforts were to be used to rouse habits of industry and effort' in the belief 'that the spiritual elevation of a large class depended on sanitary reform'[8] – in other words on decent housing. By 1874 Octavia was managing fifteen properties on behalf of other owners. These were mostly rehabilitated buildings rather than large new blocks since she had found that familiar streets were where most people wanted to stay, provided that the dwellings were in habitable condition. In a letter to her fellow workers she summarized her methods:

Each block is placed by me under a separate volunteer worker, who has the duty of collecting, superintending cleaning, keeping accounts, advising as to repairs and improvements, and choice of tenants. [The volunteers offer] all personal help that can be given to the tenants without destroying their independence, such as helping them to find work, collecting their savings, supplying them with flowers, teaching them to grow plants, arranging happy amusements for them, and in every way helping them to help themselves.[9]

The volunteers were in due course supplemented by paid workers. Octavia Hill's lasting achievement was to focus public attention on the appalling housing of the time and to demonstrate that a new profession of trained housing managers which she initiated could help to improve the condition not only of the properties but also of the people living in them.

In 1875, ten years after buying her first tenement, Octavia took a leading part in the attempt to prevent building on Swiss Cottage Fields, the slope running up to Hampstead village. In this she failed. But the campaign brought her into contact with Robert Hunter and alerted her to the rapid loss of open land in and around London. For the rest of her life she devoted much of her energy to the preservation of open spaces, from small disused burial grounds, which would provide 'outdoor sitting rooms' for the poor, to extensive stretches of the Surrey hills and the Cumbrian valleys. Her belief that 'the need of quiet, the need of air, the need of exercise, and . . . the sight of sky and of things growing seem human needs, common to all men'[10] was derived from her own experience when as a girl she had been suddenly removed from the country to central London. Later she sought recovery from her not infrequent periods of nervous exhaustion in the Lake District. And as she grew older she spent more and more time at the cottage in Kent which she shared with Harriot Yorke, her close companion.

By the 1880s Octavia's reputation as a social reformer was well established. She knew or had worked with most of the leading figures in social welfare in Victorian England. In 1884 her sister Miranda wrote to a friend, 'It *has* come to a point! – when two peers and a cabinet minister call and consult her in one week. She had Fawcett here yesterday, Lord Wemyss the day before to ask what he should say in the House of Lords and the Duke of Westminster on Wednesday to ask what the Prince of Wales could do in the matter [of open spaces].'[11] Her influence was immense and, whether she provoked agreement or disagreement, she moulded opinion. One visitor remembers her vividly: 'On top of that dumpy, oddly dressed figure there was a head of such massiveness with so broad a brow, and a glance so steady, as only greatness fortified by some consciousness of greatness could have produced.'[12] She was possessed of an iron will. 'She might survey her tenants benignly on a Sunday afternoon, but on Monday morning they had to pay the rent.'[13] In her work for the National Trust this combination of leadership and determination worked to such effect that every appeal she launched to buy land or acquire an historic building was successful. She continued to play an important part in the Trust's fund-raising and shaping of policy until her death in 1912.

Ruskin again intervened at a crucial juncture in Octavia Hill's life when in 1874 he advised Hardwicke Rawnsley, the third of the Trust's founding trio, to consult her as 'the best lady abbess you can have for London work'. Rawnsley had come down from Oxford and was helping at a lodging house for vagrants in the parish of St Mary's in Soho.

Rawnsley was born on 18 September 1851, one of eleven children. He spent most of his childhood in Lincolnshire, where his father took on the living at Halton Holgate near Skegness when his grandfather retired. His parents were close friends of the Tennysons: as a boy Hardwicke came to admire Alfred, the future poet laureate, about whom he later wrote in

Memories of the Tennysons. At the age of eleven he was sent to Uppingham, where Edward Thring, the educational reformer, was the inspiring headmaster. A contemporary at school remembers Rawnsley as 'country bred, with keen powers of observation and a retentive memory . . . an expert on the notes of birds and the ways of woodland creatures'.[14] It was Thring who introduced Rawnsley to the Lake District, taking him on family holidays to Grasmere, and who instilled the creed, 'I hold that man to be pre-eminently a man of God, who in this diseased modern world helps the weak to healthy enjoyment.'[15] The Lakes, which dominated Rawnsley's life, gained an extra appeal from his love for the Lakeland poets, especially Wordsworth. He wrote verse on almost any occasion, but the Tennysonian gift had not rubbed off, and verse to him was more a habit than a vocation.

In 1870 Rawnsley went up to Oxford, an eager, high-spirited Balliol freshman. He did not immediately impress the Master and classical scholar, Dr Jowett, but he found a more sympathetic mentor in Ruskin, then Slade Professor of Art. Rawnsley did not distinguish himself academically, emerging after a change of course with a third-class degree in natural science. He toyed with the idea of becoming a doctor before deciding to take up holy orders.

After a short time in London and two years in charge of the Clifton College Mission in Bristol, where he found himself 'half-parson, half-policeman' and was known as 'an energetic, public-spirited, dancing, laughter-loving curate'[16] he was offered the small parish of Wray on Lake Windermere. In the same year he married Edith Fletcher. Before the wedding, in response to an anxious enquiry from his future mother-in-law, his former headmaster had sent a reassuring and perceptive testimonial: 'He is so far from commonplace, so original, so full of strange power, that I find it impossible to form the same kind of absolute judgement on his future that I could do in less exceptional cases, but I have every confidence that the outcome will be good.' His wife shared with him a love of nature and the Lakes, and was impressed by his 'mine of energy and enthusiasm, ready to be fired at any moment by an easily kindled spark of interest'.[17]

It soon became clear that the solitude of Wray, however beautiful, was too remote for someone so restless and active without occasional sorties further afield. In January 1879 Rawnsley took Edith on a six-month tour of the Middle East, visiting Cairo, Gaza, Petra and Beirut. This was the first of many foreign travels, which resulted in several books, including *Notes for the Nile* and *Sonnets in Switzerland and Italy*. On one memorable visit he was commissioned to report on the Tsar's coronation in Moscow, a tribute to his descriptive powers.

In 1883 he became vicar of Crosthwaite, the parish church of Keswick, which the then Bishop of Carlisle described as being 'as near to Heaven as anything in this world can be'. The vicarage, with views over Derwentwater, was to be his home for thirty-four years and the base for his consuming passion – to save the Lake District from desecration.

Rawnsley had taken up the fight soon after moving to Wray, opposing the construction of a reservoir at Thirlmere. In this he was unsuccessful, despite the eloquent pleas of Octavia Hill in a volume of essays, *Our Common Land*. But a few years later he was able to secure the withdrawal of a Bill to build a railway along the western side of Derwentwater. Realizing that this would not be the last of such threats, he launched the Lake District Defence Society at a meeting presided over by Matthew Arnold, the object being to protect the Lake District 'from those injurious encroachments upon its scenery which are from time to time attempted from purely commercial or speculative motives without regard to its claims as a national recreation ground'.[18] These claims were recognized sixty-eight years later when the Lake District was designated one of the first National Parks.

In 1885 public indignation was aroused when two land-owners without warning barred footpaths near Keswick which gave access to Skiddaw. This was the sort of crisis that roused Rawnsley to write: 'It is a public duty to make every reasonable effort to preserve ancient rights of way in order to hand on to succeeding generations our pleasant byways as well as our necessary but sometimes unpleasant highways.'[19] Rawnsley's 'reasonable efforts' involved reviving the moribund Keswick and District Footpath Association and leading demonstrators, on one occasion numbering more than two thousand, on walks over the banned paths. At his request Octavia Hill and Robert Hunter launched their own appeal, Robert Hunter writing, 'Unless the attempts to shut the Keswick paths are strenuously resisted, the right of access to the mountain tops . . . in the Lake District will be imperilled.'[20] After a court case the paths were finally declared open to the public in 1888.

Rawnsley's second wife and biographer notes that:

Hardwicke had gradually come to be looked upon as the watchdog of the Lake District, and successive schemes which he felt likely to injure its natural beauty were continually the means of spurring him to action. No matter, great or small, which affected the Lake Country or the wellbeing of those who dwelt in it, failed to rouse his interest – his eager support if the project seemed to him to be good, or his tireless opposition if the scheme did not commend itself to him.[21]

For seven years he was able to advance this interest as an elected member of the newly created Cumberland County Council and as chairman of its Highways and Bridges Committee. But he lost his seat in 1895, partly because of his opposition to the granting of a liquor licence to the proprietors of the Keswick Pavilion.

In 1898 he was offered the Bishopric of Madagascar but was persuaded not to accept by many friends, including Octavia Hill, who wrote:

I have the profoundest honour for the men who go to the ends of the earth . . . But it is not everyone's duty . . . And what of your work you have here in England! It almost seems unique. I am sure that the National Trust owes, and must owe, much of its special character to your special influence. If it is to gather in the givers, if it is to seize upon the important opportunities for good, if it is to retain some element of poetry and of hope, it seems to me as if it must depend on you. There is no one among us who has so wide and appreciative an outlook, who brings in so many harmoniously, who can shew the poetry of a place so vividly. I really should feel most discouraged about the Trust if you were to leave us.[22]

Instead, at the suggestion of the Duke of Westminster, he substituted for semi-permanent exile to Madagascar a temporary visit to America, where he endeavoured to bring the aims and work of the National Trust before the American public.

Apart from his work for the Trust and in his parish, Rawnsley directed his prodigious energy to a bewildering range of causes. He spoke all over the country on behalf of the Armenian Relief Fund, raised a number of memorials including one to Caedmon, England's first Christian poet, campaigned against 'pernicious literature' and was one of three men responsible for the lighting of 2548 bonfires throughout Britain to celebrate Queen Victoria's Diamond Jubilee.

A continuing interest was education. In his Oxford days he had written, 'If it were ever my lot to teach children . . . I would interest them in the things of nature daily around them, and this because it would lighten their hours of toil and make them go down happy to their graves.'[23] This he did on his frequent visits to local schools. Soon after moving to Crosthwaite he and his wife started evening classes in drawing, carving and metalwork. Encouraged by Ruskin, who was then living partly at Brantwood on Lake Coniston, the classes developed into the Keswick School of Industrial Arts. Rawnsley defined one of its objectives as: 'To make it felt that hand-work did really allow the expression of a man's soul and self, and so was worth doing for its own sake, and worth purchasing even at some cost to the buyer.' Later Rawnsley was the prime mover in founding the Keswick High School and served as chairman of the county council Committee for Secondary and Higher Education.

In 1909 he was appointed a canon of Carlisle Cathedral. In congratulating him, John Bailey, later chairman of the Trust, wrote, 'I suppose you are the hardest unpaid worker in these islands, and those who have anything to do with the causes to which you devote yourself will all rejoice when they see your appointment. We of the National Trust, at any rate, know what the country owes to you, and are glad the Bishop of Carlisle should show that he knows it too.'[24]

He continued his active life until the death of his wife Edith in 1916. The following year he retired and moved to Allan Bank, a house set high on a spur above Grasmere where Wordsworth had once lived. By the summer of 1918 he had recovered enough to ask Eleanor Simpson, who had long been a friend of his and of his first wife, to marry him. They went on a tour of National Trust properties which resulted in one of his last books, *A Nation's Heritage*, published in 1920, the year he died. The Master of Trinity, Dr Montagu Butler, who was on the Council of the Trust, composed this inscription beneath a portrait of Rawnsley: 'Poet and lover of his country: of its beauty – whether of mountain or lake, wood or path – the indefatigable defender.'[25]

The contribution to the Trust made by the three founders was summed up by Nigel Bond, secretary of the Trust from 1901 to 1911, on an occasion held to mark Octavia Hill's centenary:

Hunter was a man of very wide vision, very long-sighted, an idealist, and in addition (a rare combination) an extremely able

lawyer . . . Canon Rawnsley was different. He did not claim
to be a business man, but he was a great enthusiast who did
not know the meaning of the word 'No' and it would have
been his way to have said, 'There is no obstacle there: go on'
and he would have given a leap and nine times out of ten
would have landed safely on the other side. Miss Hill, like Sir
Robert Hunter, would see the obstacle: she would not bother
to find a way round, but she would say, 'This is an obstacle
we have got to face, and we will build a ladder and we will
start on a sound foundation, climb up, get over, and go
down the other side.'[26]

Hugh Lupus Grosvenor, first Duke of Westminster, who
became the Trust's president, was one of the wealthiest and
most influential men in England, seeing himself 'not so much a
private millionaire as the head of a great public institution or
trust'.[27] Having sat as Member of Parliament for Chester for
twenty-two years, he succeeded his father as third Marquess
before being the last man to be elevated to a non-royal
dukedom. He had been a general supporter of Gladstone and
was a man of broad principles, although he placed what he
believed to be the national interest above party loyalties when
he played a leading role in defeating the Reform Bill of 1866
and in the break over Home Rule in 1886. His wealth was
derived from 600 prime acres in central London and more than
30,000 acres in Cheshire, where he spent much of the year at
Eaton Hall. His other houses included Grosvenor House in
London and Cliveden in Buckinghamshire. However, with his
generous support of a number of charities and a desire to see
his younger sons and daughters provided for, in 1893 he
decided to sell Cliveden to the William Waldorf Astor of his
generation, who brought a large slice of his family's Manhattan
wealth across the Atlantic and became the first Viscount Astor.
(His son, the second Viscount Astor and the husband of Nancy
Astor, gave the house to the Trust in 1942.) Among other

causes the Duke campaigned against cruelty to animals and for the provision of drinking fountains and cattle troughs. His enthusiasm for racing (he won the Derby five times) led the writer of *The Times* obituary to note, 'he could pass from the racecourse to a missionary meeting without incurring the censure of even the strictest'.[28] He was the ideal choice to mobilize support for the new Trust.

Westminster was president of several London hospitals, and in 1870 he joined Octavia Hill on the Committee of the Charity Organization Society. This had followed the many voluntary societies set up earlier in the century to promote better conditions for people living in crowded terraces without proper sanitation or uncontaminated water, let alone schools, parks or other amenities. Then came the first campaigning organizations for what we would now term our 'heritage': the Commons Preservation Society (CPS), founded in 1865, and the Society for the Protection of Ancient Buildings (SPAB), founded in 1877. These were the main head-springs of the National Trust.

William Morris, designer, writer and social reformer, was the SPAB's founder and leading spirit, but its intellectual inspiration came from John Ruskin. Ruskin was a quintessential Victorian. Born in 1819, the same year as the future Queen, and dying in 1900, a few months before her, his writings were imbued with concern for moral issues and made memorable by the richness of his language. Never a systematic thinker – indeed his ideas were often contradictory – he was consistently a lover of nature and natural beauty. This love of nature, instilled in childhood and fostered during journeys abroad with his parents, dominated his life. During these journeys he drew and sketched every day, becoming a skilled draughtsman.

In 1869 he was appointed the first Slade Professor at Oxford and he lectured there intermittently until he resigned through ill-health eight years later. Rawnsley recalled his magnetic power:

Ruskin got, in quite an astonishing manner, at the heart of the undergraduates. One looks back to the crowded audiences at his Oxford lectures, to the delightful breakfasts with 'the Master' [used in the sense of 'Maître', not Head of House], to the new experiences, with their fruit for life, of being roadmakers for the Hinksey poor under the direct encouragement and personal supervision of the Slade professor.[29]

Despite the mockery of their contemporaries Ruskin persuaded a number of young men to help mend the road to the village of Hinksey just outside Oxford. Rawnsley noted that many of the 'diggers' owed their interest in social movements 'to the spirit gained from Ruskin the roadmaker'.

Ruskin's love of natural beauty and the picturesque led him to dislike any form of restoration, whether of architecture, decoration or paintings. He became one of the first and most influential guardians of past work in its original form at the time when his reputation as an arbiter on artistic matters was supreme. He believed that the preservation of old buildings ought not to be a question of expediency: 'We have no right whatever to touch them. They are not ours. They belong, partly to those who built them, and partly to all generations of mankind who are to follow us.'[30] These words, reprinted in the SPAB's manifesto, express one of the themes of *Seven Lamps of Architecture*, which Ruskin published in 1849 at the age of thirty when he was at the height of his intellectual powers. He argued that the historic and artistic value of old buildings lay in their actual surface, the surface providing visible evidence of age and the tool marks reflecting the craftsmen who made them. 'Do not let us talk of restoration, the thing is a lie from beginning to end. That spirit which is given only by the hand and eye of the workman, can never be recalled.' But in one of his swings of mood from elation to depression he wrote later, 'I never intended to have republished this book which

16

has become the most useless I ever wrote: the buildings it describes with so much delight being now either knocked down or scraped and patched up into smugness and smoothness more tragic than uttermost ruin.'[31]

In this passage he was deploring the practice of Sir George Gilbert Scott and other leading mid-Victorian architects who, when called on to repair medieval buildings, 'restored' them to what they conceived to be the purest form of Gothic style rather than limiting the works to those needed to preserve the fabric. Their attitude was well described by Viollet-le-Duc, who was an equally active restorer in France and who wrote, 'To restore a building is not to preserve it, to repair or rebuild it, it is to reinstate it in a condition of completeness which could never have existed at any given time.'[32]

By 1855 William Morris was attacking the methods used to restore Ely Cathedral, but it was not until 1877 that he was moved to found the SPAB. In a letter to the *Athenaeum* that March he wrote:

> My eye just now caught the word restoration in the morning paper, and, on looking closer, I saw that this time it is nothing less than the Minster of Tewkesbury that is to be destroyed by Sir Gilbert Scott. Is it altogether too late to do something to save it – it and whatever else of beautiful or historical interest is still left us on the sites of the ancient buildings we were once so famous for? Would it not be of some use once and for all, and with the least delay possible, to set on foot an association for the purpose of watching over and protecting these relics . . .?[33]

Many artists, but initially only a handful of architects, rallied to Morris's society, of which he was secretary and the leading spirit. By the time the National Trust came into being the antagonism aroused by the SPAB in the Church of England and the architectural profession was fading and the Society was able to devote some attention to buildings threatened by neglect rather than by excessive restoration. Not having the power to hold buildings for preservation itself, the SPAB saw in the Trust a possible source of help and quickly gave its support to the new organization.

It was the Commons Preservation Society, more concerned with practical legal questions than with philosophical and aesthetic ideas, that brought together Robert Hunter and Octavia Hill. The CPS had been founded by George Shaw-Lefevre, later Lord Eversley. He was born in 1831, the son of a Speaker of the House of Commons, and was himself a Member of Parliament from 1865 and a member of successive Liberal governments up to 1895. But his interest in open spaces was to take up much of his long life (he did not die until 1928) and as First Commissioner of Works he was able to secure the opening of Hampton Court Park and Kew Gardens. Many influential politicians and social reformers supported the CPS, including John Stuart Mill, Thomas Huxley, Sir William Harcourt (later the Chancellor of the Exchequer with whom the National Trust was to negotiate over the question of death duties) and the Duke of Westminster.

A common is a misleading term to the extent that it is not a piece of land where anyone may roam. It is land which usually belongs to one person, the lord of the manor, over which other people, commoners, have certain rights, for example to graze cattle. The system of land tenure dates back to the Middle Ages, as does the procedure enabling the lord of the manor to enclose common land and use it for his own purposes. Enclosure took place sporadically until it gathered momentum in the mid-eighteenth century as landowners realized that only by enclosing land could they introduce alternative crops, new machinery and better stock breeding. Private Enclosure Acts empowered landowners to enclose their common land, often literally with a thorn hedge or wooden fence, while the landless peasants were left with nowhere to graze their cattle. By the mid-nineteenth century millions of acres of commons had become productive farmland and the remaining commons were unevenly distributed: there was little in the midlands, but in Surrey nearly every parish had its common and there were large areas of upland commons in the north and west. In the south of England commons were becoming valuable building sites just at the time when their importance for the recreation of people living in densely populated areas was being recognized. Many commons were neglected, used as rubbish dumps or treated as wastes, a term by which they were often known. Writing in the 1820s, William Cobbett, champion of radicalism and author of *Rural Rides*, noted, 'Wastes indeed! Was it a waste when a hundred healthy boys and girls were playing there on a Sunday afternoon instead of creeping about in filth in the alleys of a town?'[34]

This point of view gained ground and in 1845 a General Enclosure Act was passed which provided that the consent of one-third of the commoners was necessary before enclosure would be granted and that, if granted, some land must be allocated for recreation and allotments. But the Act had little effect: between 1845 and 1864 more than 614,000 acres of commons were enclosed and only 4000 were set aside for the benefit of the commoners. Then in 1865, in response to a threat to build on Putney Heath, a Parliamentary Committee was set up to 'inquire into the best means of preserving for public use the Forests, Commons and Open Spaces in and around the Metropolis'. The Metropolitan Board of Works proposed a compromise: the interests of the owners should be bought up, parts of the commons should be built on and the rest should be left for the public to enjoy. Shaw-Lefevre, who was a member of the Committee, rejected this view on the grounds that the commoners had rights to entire commons and not just to parts of them and that these rights still existed under the 1235 Statute of Merton, which was intended to improve grazing without diminishing the benefits of commons. The majority of the Committee supported Shaw-Lefevre and their report led to the Metropolitan Commons Act of 1866. The Act rendered enclosure of commons virtually impossible within a fifteen-mile radius of Charing Cross, though often only after protracted legal proceedings to defeat the would-be enclosers. It was not until some years later that legislation was passed to protect commons outside the London area. The hiatus had disadvantages, for the threat of restrictive legislation propelled some landowners towards enclosure or illegal fencing.

Soon after its formation in 1865 the CPS began a series of lengthy but ultimately successful battles to preserve open spaces, such as Hampstead Heath and Epping Forest within the metropolitan area and Berkhamsted Common and the Malvern Hills outside. But for these victories London would not now have its commons and woodlands to complement its royal parks. It was Epping Forest, out of which arose the longest and most complex of the Society's cases, that made the name of Robert Hunter.

Hunter took the essential first step towards founding the National Trust in September 1884 when, in a speech to the National Association of Social Science in Birmingham entitled 'A Suggestion for the Better Preservation of Open Spaces', he outlined his idea for a company with the power to acquire and hold land and buildings. Rather than working primarily for profit the company would have as its objective the protection of the public interest. Hunter had been made aware of the need for such a body when Octavia Hill had consulted him about the house and grounds of John Evelyn, the seventeenth-century diarist, in Deptford, south-east London. The house had remained in the Evelyn family and his descendant wished to present it to the Metropolitan Board of Works. But even Hunter could find no means by which the Board could accept the house, and the offer lapsed.

Octavia Hill was enthusiastic about the proposed company, which she hoped would be ready if and when some other important property came forward. In February 1885 she wrote to Hunter:

> A short expressive name is difficult to find for the new company. What do you think of the Commons and Gardens Trust? I do not know that I am right in thinking that it would be called a Trust. But if it would, I think it might be better than 'Company' – you will do better, I believe, to bring forward its benevolent than its commercial character. People do like charity when a little money goes a long way because of good commercial management.

At the top of this letter Robert Hunter has written in pencil, 'The National Trust?'[35]

The preparations for setting up the Trust were at first strangely slow. Another ten years were to elapse before it was formally accepted by the Board of Trade, perhaps because both Octavia Hill and Robert Hunter were immersed in other public work. They had also to contend with the unexpected opposition of Shaw-Lefevre, who rightly feared that the existence of the Trust would diminish the role and funds of the CPS. Octavia Hill in consulting Hunter about a possible chairman suggested the name of James Bryce, who had been influential in the CPS and as a respected Member of Parliament had offered to help establish the Trust. In fact Hunter became the first chairman.

A draft memorandum of association was drawn up with objectives very similar to those of the Kyrle Society, which had been founded by Octavia Hill's sister Miranda 'for the Diffusion of Beauty' and included among its objectives the use of buildings 'for the purposes of recreation or instruction' and the securing of burial grounds as open spaces. Its decorative branch recruited William Morris as a lecturer and its open-spaces sub-committee, where it was most effective, had Hunter as chairman and honorary legal adviser. In certain respects the Kyrle Society foreshadowed the National Trust but, without focus or structure, it failed to attract enough committed members and it foundered as the Hill sisters grew older.

A more direct influence came from across the Atlantic. Hunter must have been well aware that the United States had given a lead in preservation not only through the establishment of the first national park but also through the work of local historical and nature conservation societies in buying up old buildings and natural sites. At his prompting, Octavia Hill wrote to Ellen Chase, an American who had worked for her in Deptford, asking for reports on the open-space movement in the USA, in particular the constitutions of the Boston Metropolitan Park Commission and the Trustees of Public Reservations in Massachusetts. The Trustees of Public Reservations had been set up in 1891 'for the purposes of acquiring, holding, arranging, maintaining and opening to the public under suitable regulations beautiful and historical places and tracts of lands'. The corporation was empowered to 'acquire and hold by grant, gift, devise, purchase or otherwise real estate such as it may deem worthy of preservation for the enjoyment of the public but not exceeding $1m'.[36] This wording clearly influenced Hunter when he came to draft the constitution for the Trust.

It was not long before pleas for assistance came from Hardwicke Rawnsley in the Lake District, where in the early 1890s a number of important properties came on to the market. Fearing that, without protection from the proposed trust, beauty spots such as the Falls of Lodore would soon be ruined, Rawnsley gave powerful support to the campaign.

Aware of the urgency, the three joined forces and wrote to likely supporters on paper headed 'National Trust for Historic Sites and Natural Scenery'. Shaw-Lefevre was persuaded to withdraw his opposition and a preliminary meeting to discuss the formation of the Trust was held at the offices of the CPS on 16 November 1893. The reports next day were enthusiastic. *The*

Times wrote, 'We see no reason why for public purposes a bit of beautiful scenery should not be the subject of a forced sale under equitable conditions just as much as a bit of ugly country for a railway.'[37] The *Daily News* described the new National Trust as 'the Commons Preservation Society in active rather than merely advisory functions . . . If we had had them earlier in our history, Mr Ruskin would have been spared the necessity of writing many a mournful page.' It continued: 'Local authorities can hardly be expected to help the public to preserve the beauty of its great pleasure grounds – their area of action and the sources from which they draw their funds are, as a rule, too contracted, and they have many claims upon their not too ample resources.'[38] A Provisional Council was formed and it was decided that the Trust be incorporated under the Joint Stock Companies Acts, and that because of its non-profit-making status the use of the word 'limited' would be omitted.

CHAPTER TWO

Early Years

WITH HUNTER, Hill, Rawnsley and Westminster, the other names on the Provisional Council made an impressive list. Among them were the Prime Minister (the Earl of Rosebery), the Marquess of Dufferin and Ava, the Earl of Carlisle, Professor Thomas Huxley, the Provost of Eton (James Hornby) and the Master of Trinity College, Cambridge (Montagu Butler). Organizations which the founders considered would be useful to the new Trust were also represented. These included the Linnean Society, the Royal Academy of Arts (Sir Frederic Leighton), the Royal Botanic Society and the Trustees of Public Reservations in Massachusetts, as well as the CPS, represented by Shaw-Lefevre, and the Kyrle Society, represented by Octavia Hill's close friend Harriot Yorke. The SPAB was not represented until the following year.

When the Provisional Council met on 16 July 1894, at Grosvenor House, Park Lane, the Duke of Westminster's London house, Octavia Hill proposed the motion that it was 'desirable to provide means by which landowners and others may be enabled to dedicate to the nation places of historic interest or natural beauty, and that for this purpose it is expedient to form a corporate body, capable of holding land and representative of national institutions and interests'. The aim, she explained, was to 'save many a lovely view or old ruin or manor house from destruction and for the everlasting delight of thousands of the people of these islands'. Robert Hunter then moved 'that this meeting approves generally of the proposed constitution of the National Trust for places of Historic Interest or Natural Beauty as explained to the meeting, and authorizes the necessary steps to be taken to procure the legal incorporation of the Trust'.[1] Hardwicke Rawnsley was appointed honorary secretary, a position he held until his death in 1920.

The day after this meeting *The Times* declared that the Trust 'aims to establish a National Gallery of natural pictures', just as the 'dormant sense of the beautiful was awakening in the mass of our population'.[2] Octavia Hill noted that 'the Trust, like St Francis of old, would be strong in poverty and like him, would appeal for gifts'.[3] *Punch* referred to it as 'The Grand National Trust', as though it were a new steeplechase, and commended so patriotic a scheme.[4] The articles of association, drawn up by Robert Hunter, were submitted to the Board of Trade, and on 12 January 1895 'The National Trust for Places of Historic Interest or Natural Beauty' was formally registered under the Companies Act, vested with the power to 'promote the permanent preservation for the benefit of the Nation of lands and tenements (including buildings) of beauty or historic interest'. When Octavia Hill had first proposed the idea of a trust to the Duke of Westminster, he wrote back in prophetic words, 'Mark my words the National Trust has a great future before it.'[5]

By the opening meeting of the twelve-member Executive Council in February 1895 the Trust had acquired its first property – Dinas Oleu, a four-and-a-half-acre cliff-top above the seaside town of Barmouth in Wales, given by Mrs Fanny Talbot. The cliff, which is reached by steep stone steps and a hidden track behind the town, looks out over the Mawddach estuary and Cardigan Bay. The site could not have been more appropriate. Not only was Mrs Talbot a close friend of Rawnsley but she had also known Ruskin, with whom she had stood upon this cliff some twenty years earlier. Octavia Hill gave it her full approval; in a letter at the time she mused, 'We have got our first piece of property, I wonder if it will be the last?'[6] Rawnsley, voicing his passion for the new Trust while staying with Mrs Talbot, had persuaded her to present the property even before the articles of association had been finalized. When he returned to Dinas Oleu in 1918 he noted, 'From no place on earth could these rough rock-dwellers have a finer view of sea or sky, or glorious glimpse of the mountains, their changing lights and shadows on either side of the broad estuary.'[7]

Mrs Talbot had no doubts about her donation:

I am grateful for this chance for I perceive your National Trust will be of the greatest use to me. I have long wanted to secure for the public for ever the enjoyment of Dinas Oleu, but I wish to put it into the custody of some society that will never vulgarize it, or prevent wild nature from having its way. I have no objection to grassy paths or to stone seats in proper places but I wish to avoid the abomination of asphalt paths and the cast-iron seats of serpent design which disfigure so largely our public parks and it appears to me that your association has been born in the nick of time.[8]

The Trust tried hard to persuade the Lord Harlech of the day to give some of his land adjoining Dinas Oleu, but there was disagreement about fencing and nothing came of it.

The first building acquired by the Trust was a perfect example of the sort that reflected the enthusiasm of Ruskin and William Morris for medieval architecture. It was a fourteenth-century clergy house in the village of Alfriston at the foot of the Sussex Downs and was bought for £10 from the Ecclesiastical Commissioners in the winter of 1896, shortly after the death of William Morris. In urgent need of repair, it had been brought to the Trust's attention by the SPAB, which did not itself have the power to hold buildings for preservation. 'In places the thatched roof was open to the sky, rain streamed through, unhindered, to the rooms below. Several walls bulged ominously; within, the staircase leading to the upper floor had entirely disappeared. It stood a forlorn relic.'[9] Nevertheless it was one of the few fourteenth-century domestic buildings left standing in southern England and this alone warranted the £350 that was needed to restore it.

With her experience of rehabilitating slum property it was natural that Octavia Hill should take charge of the repairs. Writing to a benefactor, she recognized that 'to let it deteriorate further would be a sort of breach of trust ... because it is ours now, given in the expectation we could preserve it. Besides all this, hope is a great factor in inspiring people to work and gift, and if our National Trust failed in these small schemes in this the opening of its work, it would throw back the future work.'[10] But she did not find it easy to raise the necessary funds – 'All my friends seem keener about beautiful open spaces ... we don't seem to reach the antiquaries and artists.'[11] The Trust was less successful in this respect than was the National Art-Collections Fund, which was founded in 1903 in order to stem the export of works of art sold from country houses mainly to American collectors and museums, and was for some years well ahead in the size of its membership.

The first balance sheet to be published, for 1896–7, demonstrates Octavia Hill's difficulty: a sum of £545 was raised for the purchase of Barras Head in Cornwall, compared with only £215 for repairing the clergy house. The Trust's total income for the year was £991. £137 of this came from annual subscriptions, not enough to cover general expenses, including the salary of the secretary, who for the first couple of years was shared with the CPS. (A factor of about 40 should be applied to indicate the increase in prices in the 1990s over 1900 prices though some items, for example average wages, have increased by more.) The report stressed that the Trust was run on the most economical lines and that most of the work was done by Rawnsley and others who received no remuneration. The Trust preferred its current expenditure to be met by a large number of small subscribers, though large donations were implicitly welcomed for new acquisitions, and the hope was expressed that in the future there might be a claim on the national purse.

The early annual reports reveal that many properties were under discussion, although the acquisitions were relatively few. The immediate response had possibly not been as enthusiastic as had been expected, despite the publication in 1897 of the Trust's first promotional pamphlet, *Its Aims and Its Works*, written by Robert Hunter. This was perceived as a policy document and outlines what the Trust was trying to achieve: 'It is the only association that, in the absence of any power under Parliamentary statute to safeguard other than prehistoric remains, can take it upon itself to preserve for posterity historic sites and buildings that may be handed to its keeping. It is thus the friend alike of historian, painter and poet.'

The pamphlet sets out the Trust's desire not only to take on property but also to support the preservation of places thought worthy in terms of historic interest or natural beauty. 'Whether it be a waterfall destroyed as in the case of Foyers, or an old bit of Sir Christopher Wren's London . . . that is threatened, or the quietude of Kynance Cove and the destruction of the rocks of the Cornish foreshore . . . that is brought under its notice . . . the Trust brings its influence to bear.' The Trust had no reason for existence except a 'purely patriotic interest in those things which in the crush of our commercial enterprise and in the poverty of landholders or in the lack of local care, run risk of passing away'.[12]

During its first fifteen years, the Trust campaigned energetically for the preservation of sites and historic buildings in different parts of the British Isles, from churches and houses near its office in London to the Wall of Antoninus north of Glasgow and the Lakes of Killarney in Ireland, which it hoped would become a national park on the American model.

Stonehenge, which was to be a continuing source of concern for the Trust, featured in its first year when the owner, Sir Edmund Antrobus, was asked if he would sanction an attempt to get help from public funds to prevent the stones from falling. He did not reply and several years later when two of the massive uprights had indeed fallen over the Trust proposed that Stonehenge be made the property of the nation and opened to the public. In the ensuing controversy the Trust sided with the CPS (which had merged with the National Footpaths Preservation Society in 1899) in their efforts to retain free access to the monument, rather than with the SPAB, who were primarily concerned with protecting the stones from damage. It was not until 1915 that agreement was reached for Stonehenge to be bought by the Office of Works.

Another imaginative but unsuccessful initiative was taken when part of Snowdon was offered for sale: the Trust asked if the owner would be willing to transfer the summit to them, but he refused. More successful (at least until the Second World War) was the opposition to a proposal to manufacture hydro-planes and to train aviators on Lake Windermere.

By 1908, although the Trust was still a very small concern, operating from one or two rooms in Victoria Street, it had acquired thirty properties, and their management required increasing attention. There was thus less time to fight for threatened sites all over the country, but until his death in 1913 Sir Robert Hunter continued to put forward imaginative proposals for legislative reform. As a preliminary he published a report on the measures taken for protecting outstanding landscapes and historic buildings on the continent and in North America. Following this study, summarized in the Trust's third annual report, he presented three draft Bills. The first concerned ancient monuments and was influential in securing the Act of 1900, which established the principle of public access to scheduled ancient monuments. The other two Bills were far ahead of their time: one provided for the government to have power to suspend the destruction of places of national interest while time was allowed to devise means of permanent acquisition (similar in intention to powers introduced in 1947); and the other, the Land Dedication Bill, provided for an owner to be exempt from death duties if he dedicated historic monuments or land to the public (similar to the measure introduced in the Finance Act of 1976).

The first effective Ancient Monuments Act, which was passed in 1913 and set up a board with professional inspectors to list monuments of national importance, was also partly due to the persistent efforts of Hunter and the Trust. Public opinion had been alerted to the dangers by Lord Curzon's dramatic intervention to save Tattershall Castle in Lincolnshire, which he later left to the Trust. In the debate running up to the Act of 1913 he commented, 'This is a country in which the idea of property has always been more sedulously cherished than in any other, but when you see that to get that Bill [Ancient Monuments Protection Act 1882] through Parliament it had to be denuded of its important features and only after many years was it passed in an almost innocuous form into law one feels almost ashamed of the reputation of one's countrymen.'[13]

Hunter was also influential in securing changes to Lloyd George's budgets of 1909 and 1910. Under the Finance Act of 1910 works of art of national importance were exempted from death duty unless sold; and the Inland Revenue was empowered to accept land and buildings instead of cash in payment of death duties. Although this power was used on only two occasions before 1945, it anticipated the National Land Fund established by Hugh Dalton as Chancellor of the Exchequer in 1946.

But Hunter's most important service to the Trust was in drafting the first National Trust Act and securing its passage through Parliament in 1907. By this time the Trust owned more than 1700 acres and it was felt that stronger powers were needed for their protection and management. The Act gave the Trust power to make by-laws, but, more important, it enabled the Trust to declare its land inalienable. This meant that property belonging to the Trust could not be acquired by government departments, local authorities or any other agency without parliamentary approval achieved by a special procedure. Future donors were given a powerful assurance of security: they could be confident that any gift of property which the Trust decided to declare inalienable would be as safe as is possible in this uncertain world (infinitely safer than if given to the government or a local authority), and this is undoubtedly one of the reasons for the Trust's success. That a small organization, still very much in its infancy, was able to secure this unique power was due to Hunter's legal skill and political acumen, helped by having four MPs and five members of the House of Lords on its Council.

Many of the Trust's early properties were, by no coincidence, concentrated in the areas where the three founders lived. The Lake District was to be of particular importance, although it was not until 1902 that the Trust made its first acquisition there. Rawnsley felt that it was a great deprivation for the public not to be able to walk along the shores of the Lakes, so when Brandelhow Park on Derwentwater came on the market he launched an appeal for the Trust to buy it. The £6500 was raised within five months but there was still some apprehension about the consequences. The annual report of that year notes, 'The experiment of opening this pleasure ground to the nation will, of course, be closely watched and commented on. It is for the people who visit it to prove that they can be trusted to use it in the right way and that they can themselves be the most jealous guardians of the order and beauty of their own possession.'[14]

Brandelhow was opened by Queen Victoria's fourth daughter, Princess Louise, who had just become president of the Trust, beginning a royal connection which has continued ever since. Octavia Hill described the scene to her mother:

I have just come back from the great opening. . . . It was very successful, very simple, and real and unconventional. The place was looking very lovely, I never saw the light more beautiful. . . . The wind was high and tore the tent to ribbons when it was being put up, but I think it really did better, because the simple red dais was under the free sky, with the great lake lying below. . . . It was very funny and primitive and the nice north country people were quite near and saw and heard all. The Princess was most friendly and kind and really did show deep and intelligent interest in the National Trust work.[15]

Princess Louise was by no means the conventional princess. Married to the Marquess of Lorne, heir to the Duke of Argyll and a Liberal MP, she was a talented sculptress, a member of the SPAB and an early patron of Edwin Lutyens. Lutyens found her 'witty and downright', but Charles Ashbee, one of the leaders of the Arts and Crafts movement who had toured America on behalf of the National Trust with Rawnsley, described her as 'guelphic and a bit conventional, not so hop-eyed as the rest of the family, but very kindly. She had the royal gift of manner which is so taking when you are bowed to and thanked . . . or your hand is gently pressed when you are given your congé.'[16] She had come to know Octavia Hill through her elder sister Princess Alice and had been patron of the Kyrle Society before becoming vice-president of the Trust at its first formal meeting. At the opening of Brandelhow the Princess asked who was going to replace the Marquess of Dufferin and Ava (the Marquess had succeeded the Duke of Westminster as president in 1900 and had died only two years later). A decision had not been taken and she immediately offered herself: 'Would there be any objection to my becoming president?'[17] Octavia Hill was delighted. Princess Louise remained in her post until her death in 1939 and made regular contributions to appeals. Among them was a donation towards the purchase of Grange Fell in memory of her brother, Edward VII, the highest point being named King's How at Rawnsley's suggestion.

After the success of the Brandelhow appeal, Gowbarrow Fell on Ullswater, which was threatened by an iron bridge and extensive building, became the next target, a more substantial one. An appeal leaflet was put out which read like a crystal ball in its prediction of the future. 'Why not nationalize the English Lake District? If men had been wise enough to realize the worth of such a haunt of ancient peace the thing might have been done. Now every year the land is more locked up in private ownership.'[18] Gowbarrow Fell, with its 750 acres of woodland and rough ground, was offered for the then large sum of £12,500. The appeal attracted 600 subscribers and Gowbarrow was opened to the public on 9 August, 1906.

From then on there was a steady stream of acquisitions in the Lake District, the years 1912–13 being the most prolific: a strip of land north of Stybarrow Crag on Ullswater just before it was built on; the 'Druids Circle' at Castlerigg near Keswick as it was about to be fenced; Borrans Field at Ambleside with its unexcavated Roman fort saved from another building threat; and Queen Adelaide's Hill overlooking the upper reaches of Windermere. During the First World War, acquisitions naturally slowed down but in 1919 a fresh appeal was launched aimed at attracting gifts of land in memory of the fallen, and the Trust's Lake District estate expanded rapidly. New properties included three small pieces of land bought to commemorate the life of Rawnsley, who died in 1920 and whose work had earned him the title 'Guardian of the Lakes'. Other memorials included Scafell Pike given by Lord Leconfield (he later gave Petworth in Sussex to the Trust) and Castle Crag in Borrowdale given by Sir William Hamer.

Octavia Hill had for many years been a regular visitor to the Lake District, especially to Derwentwater, but it was round the High Weald in Kent that she concentrated her efforts to acquire land for the Trust. The annual report of 1904 records the gift of Crockham Hill:

> Like the Terrace owned by the Trust on Toys Hill, it thus carries out the suggestion made by the Council some years ago that the presentation of such spots to a body which is pledged to retain their natural beauty as far as possible is a fitting mode of perpetuating the memory of those who in their lives have shewn at once an appreciation of Nature and a desire to make a permanent gift to those who have no beautiful parks or gardens of their own, and whose lives are spent in the gloomier and more sordid surroundings of a great city.[19]

Octavia Hill was at last fulfilling her ambitions to acquire countryside near London for the benefit of those who rarely saw it. It brought her great pleasure that the Trust owned the three most prominent points near her home. 'Each forms a vantage ground for looking over (what Rudyard Kipling calls) "the blue goodness of the Weald".'[20]

About thirty miles west of Crockham Hill, Robert Hunter was raising funds to buy land for the Trust near his own home at Haslemere in Surrey. Many of his neighbours shared his concern about the speed with which the area was being swallowed up by suburbia. When 750 acres of Hindhead Common, including the dramatic landscape known as the Devil's Punch Bowl, came on the market in 1906 Hunter formed a local fund-raising committee with the help of Lawrence Chubb, secretary of the CPS. The money was soon found and Hunter became chairman of a local committee of management responsible to head office, an example followed at a number of other properties. Lawrence Chubb had been the Trust's first salaried secretary, but after a year he had moved to the CPS, where his real interests lay and which was then an established and more influential organization. During his long service as an energetic campaigning secretary up to his death in 1948 he was instrumental in raising funds to buy a number of commons and other open spaces which were then transferred to the Trust's ownership.

The early buildings acquired by the Trust reflect the affection felt by the founders and the small group of early members for small, vernacular buildings. Octavia Hill described them as:

> our small houses, steep in roof and gable, mellowed with the colour of ages, picturesque in outline, rich in memories of England as our ancestors knew it. Alfriston pre-reformation clergy house, nestled below the Sussex Downs; Long Crendon Court House, used since the time of Henry V, standing at the end of the long street of a needle-making village of Oxfordshire; the old post office at Tintagel, a picturesque 14th century cottage. . . . There was nothing great about them, nothing very striking, only quaint picturesque out-of-the-world places greeting the eye with a sense of repose.[21]

The SPAB continued to work closely with the Trust, referring buildings to it, advising on (and occasionally criticizing) the repairs, and in some cases helping to raise funds.

In 1907 the Trust became more ambitious and took on its first large country house, Barrington Court in Somerset. It was an extraordinarily rash venture. Three years earlier *Country Life* had published an article describing the house with its golden stone, high gables and spiral chimneys as 'the final and most free expression of pure Gothic design as applied to secular buildings'. But the article continued: 'Now fallen from its high estate, almost a hollow shell, yet a shell wrought with the finest skill, fashioned in the noblest manner, the house raises its speechless voice to claim the care of some regenerating hand.'[22] Perhaps it was this appeal that inspired Miss J. L. Woodward to buy the house for the Trust on condition that £1000 was provided for repairs. This sum could never have begun to meet the bill, but soon after the First World War, when the house was deteriorating further, Colonel A. A. Lyle

took on a long lease in order to provide a home for the panelling and fittings which he had been collecting from derelict
houses. He engaged two SPAB-approved architects to carry
out the restoration and to convert the seventeenth-century
stables into a house for himself, also making a garden with the
advice of Gertrude Jekyll. Thirty years after the house had
been accepted, Nigel Bond, who had been the Trust's secretary
at the time and was later to be vice-chairman, was heard to say
when discussing a property, 'We cannot possibly take it on,
remember Barrington.'[23] It was the only occasion before 1931
that the Trust strayed beyond the acquisition of open spaces,
ancient monuments and modest buildings.

Two sites acquired in the early years have since been recognized as nature reserves of international importance. The first
was Wicken Fen in Cambridgeshire, described in the annual
report as 'almost the last remnant of the primeval fenland of
East Anglia and of special interest to entomologists and botan-
ists on account of the rare insects and plants found there'. The
second was Blakeney Point, a shingle spit off the north Norfolk
coast and a haven for many species of seabirds and geese.
Generous contributions were made towards the purchase of
these two sites by Nathaniel Charles Rothschild, an enthusiastic naturalist. He had joined the Trust's Council in 1910 in the
hope that it would pursue energetically the objective set out in
section 4 of the Act of 1907. This states, 'The National Trust
shall be established . . . as regards lands for the preservation
(so far as practicable) of their natural aspect, features and
animal and plant life.'[24] This seemed to offer the opportunity
of promoting his concept of conservation, which has been
described by his daughter Miriam Rothschild, herself a distinguished biologist:

> He realized the importance of protecting and preserving the
> habitat and special biotypes rather than the individual rare
> species threatened with extinction. . . . He was convinced that
> sound conservation policy could only be based on adequate

surveys, not only of the flora and fauna of a potential reserve, but [of] the habitat in relation to the environment as a whole.[25]

A man of many talents, he had an encyclopaedic knowledge of natural habitats from the virgin forests of Hungary to the Egyptian desert, and although much of his time was spent in the family bank, he became the world's foremost authority on fleas. His early death came as a tragic loss. 'Charles was the dependable one and Charles did his duty, and even did it with distinction but at heavy personal cost. Naturally a depressive, the unwanted responsibilities of the bank and being *de facto* head of the family must have seemed almost unbearable, and when to this burden was added mortal illness, the load became too much to bear.'[26] In 1923 at Ashton Wold, Northamptonshire, to which he had first been attracted by its rare butterflies, Charles took his own life.

Charles Rothschild became frustrated with the lack of government support and impatient with the National Trust. Such was his irritation with the Trust which he described as 'haphazard and careless', that in May 1912 he set up the Society for the Promotion of Nature Reserves, which has since developed into the Royal Society for Nature Conservation. The objectives of the new society were outlined by Rothschild in its minute book: 'To urge by means of the press, by personal efforts and correspondence with local societies and individuals the desirability of preserving in perpetuity sites suitable for nature reserves, which sites are to be handed over to the National Trust under special conditions.'[27]

By 1915 the first major task of the society had been completed and a list of 284 important sites had been delivered to the Board of Agriculture. However, given the Board's preoccupation with food production during the war the list was ignored, and afterwards it was not thought feasible to preserve wildlife on islands in the wider countryside, Rothschild's ideas being met, if not with open amusement, then with incomprehension.

Charles Rothschild was not to be the last member of his family who was a generous benefactor to the Trust. After the Second World War his cousin Anthony de Rothschild gave Ascott with its collection of paintings, furniture and porcelain, and his more distant relative James de Rothschild bequeathed Waddesdon Manor with its magnificent works of art.

Individual benefactors were still a rarity. One exception was Miss Rosalie Chichester, who in 1909 gave fifty acres at Morte Point on the North Devon coast in memory of her parents, a place 'full of wild charm surrounded on three sides by the sea . . . with a dread beauty of its own'.[28] The donation was made with the specific request that the land should be left in its natural state, a point recognized in the annual report, which congratulated Miss Chichester on resisting the temptation to turn her property into profit by building a hotel, as had been done at nearby Woolacombe Sands.

Rosalie Chichester, the only child of a flamboyant and extravagant father, had inherited Arlington Court when she was fifteen. A great traveller, she had sailed around the world twice with her father. She was to spend much of her solitary life filling Arlington with artefacts collected from her extensive travels. A house parlourman, Jan Newman, who had to spend his nights in the kennels as no men were allowed to sleep in the house, describes the scene: 'The Hall enthralled me, with its great cases of stuffed birds, butterflies and albatross, a kangaroo and a large bear; while on top of the cases were sets of Famille Rose china, great dishes and vases decorated with animals and insects.'[29] As she grew older Miss Chichester became increasingly concerned with protecting wildlife: 'I am anxious to leave the National Trust a large piece of land, chiefly woods, as a Nature Reserve and also as a "beauty spot".'[30] Later she erected an eight-mile, eight-foot-high iron fence around her land in order to exclude the hunt and hounds and when she died in 1949, bequeathing her entire 2780-acre estate and house, Arlington Court, to the Trust, she asked that hunting should not be permitted.

The three founders continued to imprint themselves on the Trust so long as they were alive, and their qualities were recognized in appreciations at the time. When Octavia Hill died in 1912 Robert Hunter wrote of her:

I had exceptional opportunities of realizing the strength which she brought to any movement which she supported. Her power of brushing aside all subsidiary details and piercing to the heart of any question, her careful choice of methods and her scrupulous regard for the rights and wishes of those who supplied her with funds and her courage and boldness in carrying through any enterprise of which she approved were as valuable a moral and intellectual stimulus to those Societies as her remarkable power of raising money was an invaluable means of supplying means of action.[31]

When Robert Hunter died a year later the obituary notices were united in their praise of this modest man, for whom no project ever seemed too difficult and no problem too intractable, who never sought recognition and was always quick to deflect praise away from himself. 'His life's work has been one long record of services of which it would be difficult to exaggerate the value. . . . He laid the whole country . . . under a debt which few of the present generation realize . . .' observed the *Spectator*. 'He was one of a gradually diminishing band of men whose hold on great truths and enthusiasm for perfectly thought out and loyally and devotedly pursued ideals made them real guides and helpers.'[32] Hardwicke Rawnsley wrote an admiring if not very poetic sonnet in his memory:

Who laboured in the restless City's roar
And gave his strength to service of the State
With Nature did his soul communicate
And loving her – he loved the people more.

Unresting ever, armed with wisdom's lore,
True knight, he stood to guard the open gate
Kept paths and common still inviolate
To bless the nation with their restful store.[33]

By the outbreak of war in 1914 the Trust had recruited 725 members and owned sixty-three properties covering 5814 acres. Twenty-eight had come by gift and thirty-four had been purchased, thirteen after successful appeals. Donations given towards expenditure on acquisitions and special projects reached £59,000, a very large increase on ten years earlier, whereas annual subscriptions still totalled only £558. The membership has been described in terms of a 'London Club who shared a pleasant community of interests [and] could be relied upon to make a generous financial contribution'.[34] Many of the members contributed time as well as money: seventy-nine enrolled as 'local corresponding members' (including fifteen in the United States) in order to help with advice and information, and others served on local committees for managing particular properties. Some members were also instrumental in persuading local field clubs and archaeological societies to affiliate to the Trust (thirty by 1914). The importance attached to help from local corresponding members and affiliated organizations is indicated by the publication of their names in the annual reports until the shortage of paper during the Second World War rendered such tribute impossible.

The work of the Trust slowed down during the First World War. Appeals were abandoned and S. H. Hamer, who had been secretary since 1911, was obliged to make economies both in staff and in the maintenance of properties. Hamer was in any case no dynamo, described by a member of the Council as 'a pleasant little man capable of turning over wheels which were already moving, but certainly not endowed with enough drive to make them revolve faster'.[35] Nor did the Earl of Plymouth, who had been a distinguished First Commissioner of Works in Balfour's government and who became chairman in 1914, give the committed and innovative direction previously provided by Robert Hunter. He frequently left the chairmanship of the Executive Committee, where the Trust's business was conducted, to his deputy John Bailey, who was to succeed him.

During the war the Trust continued to receive gifts of property, including two dovecotes and limited areas of land, some presented as memorials. A larger and unexpected acquisition was Quebec House, Westerham, Kent, where General Wolfe had spent his early years. This was given with an endowment by Mrs J. B. Learmont, a Canadian.

The most significant acquisition of land, which more than doubled the area protected by the Trust, was not a freehold but a 500-year lease of over 7000 acres of Exmoor, 'one of the most beautiful pieces of wild country to be found in England'. It came from Sir Thomas Dyke Acland in agreement with his nephew and heir, the Liberal MP Francis Acland, who welcomed the decision; 'It's really a magnificent stretch of country. Jolly to think it'll all be national.'[36] The latter's son, Sir Richard Acland, demonstrated that he shared the same outlook when twenty-seven years later he gave the largest area of land ever received by the Trust as a gift during an owner's lifetime. Lord Plymouth, in a letter to *The Times*, explained that Sir Thomas and his successors would continue to enjoy the rents and all the ordinary rights of an owner except that they would have no power to develop the estate for building and the Trust would obtain 'such control over the exercise of [the owner's] other powers as may be necessary to preserve the property, so far as possible, in its present beauty and natural condition'. He added, 'The new property will not be a source of income to the Trust. But we believe that in agreeing to accept something less than ownership, the Trust is finding a new and useful means of discharging its duty to the nation.'[37] This move foreshadowed the wider use of covenants under

41

which land may not be developed or the exterior of an historic building altered without the agreement of the Trust. Covenants may be enforced in perpetuity and against succeeding owners and are a means of giving some protection without a transfer of ownership.

The Trust ceased to campaign until the end of the war was in sight, but when common land was included by the Ministry of Reconstruction in plans for large-scale conifer afforestation it felt obliged to protest. Reporting on the result of a deputation to the President of the Board of Agriculture, the annual report for 1918–19 welcomed his assurance that he 'could not conceive it possible that National Trust property would be interfered with under this or other government schemes', but regretted that he did not embrace the general issue.

In 1919 the Executive Committee discussed 'how far the Trust should engage in work of a "militant" character as opposed to its functions as a holding body', but did not reach a conclusion. Without the wider vision of Octavia Hill and Robert Hunter, the Trust was beginning to concentrate on the management of its properties and the acquisition of new ones. At this meeting John Bailey suggested that 'the time had come when the Trust should arrive at a clear understanding as to what properties it should or should not take under its charge'. The Committee concluded that there should be a report on '(a) the possible advisability of adding to the existing staff and, if so, to what extent and how the cost should be met and (b) the possibility of laying down some principle as to the definition of the terms "historic interest" and "natural beauty".'[38] The first conclusion was gradually implemented. The second has proved elusive.

CHAPTER THREE

Urban Sprawl

SOON AFTER the end of the First World War the Trust issued an appeal for gifts to commemorate the men who had lost their lives: no more fitting memorial could be found than 'some open space, some hilltop commanding beautiful views, some waterfall or sea cliff'. The most notable response came in the Lake District, where the Fell and Rock Climbing Club gave Great Gable and the nearby fells. 'What a gift it is!' said Francis Acland when receiving the deeds on behalf of the Trust. 'Imagine yourself on Black Sail, looking east. One hundred yards or so above the pass one of the areas begins and takes in Kirk Fell, Great Gable and Green Gable, passing north to Base Brown and Brandreth . . . round east and north by Great End (a very fine hill), Sprinkling Tarn, Allen Crags . . . Glaramara and north to Seathwaite Fell. What names to conjure with!'[1]

As the troops were demobilized they looked forward to 'homes fit for heroes' and a return to the jobs they had left. But the homes did not materialize and many of the jobs disappeared when recession followed the short post-war boom. More lasting were the social effects of the war in bringing about greater equality between the classes and the sexes.

The sons of the landed classes suffered a particularly high level of casualties. 'In the retreat from Mons and the first battle of Ypres perished the flower of the British Aristrocracy,' noted one historian.[2] Those who returned came back to find that their estates were threatened by higher taxation on large incomes and still more by the raising of death duties in the Budget of 1919 to 40 per cent on estates worth more than

43

£2 million. The spring of 1919 brought an avalanche of land sales. But the sales were due not only to taxation. Owners had come to realize that they could do better by selling land and reinvesting elsewhere and many felt that the low return on the capital value of land imposed a sacrifice no longer outweighed by social advantages. It has been estimated that between the end of the war and 1921 over a quarter of England changed hands, a scale of transfer which had not occurred since the dissolution of the monasteries four centuries earlier. The sales resulted in the dispersal of a large part of the great estates and a significant increase in the proportion of land farmed by owner occupiers, which rose from 11 per cent in 1914 to 36 per cent in 1927. Higher taxation could not be offset by higher income from the land. In 1921 the Corn Production Act effectively wiped out the subsidy on wheat and set off a renewed depression in agriculture from which farmers did not begin to recover until the Second World War.

For many the reaction to these troubles was to believe that pre-war England had been 'mellow, dignified, creeper-clad and bathed in perpetual sunshine'.[3] People dreamed of living in large houses in the golden summer, such as those evoked by L. P. Hartley in his novel *The Go-Between*. The 1920s brought with them a cult for the countryside which was articulated by Stanley Baldwin, Conservative Prime Minister three times between 1923 and 1937. Although he was the son of an industrialist and spent the first twenty years of his working life in the family iron and steel firm, he liked to see himself as a countryman. He helped the Trust with fund-raising when it was still a small and little-known organization. He extolled the unchanging rural life, describing his vision of England in evocative language:

> the tinkle of the hammer on the anvil in the country smithy, the corncrake on a dewy morning, the sound of the scythe against the whetstone, and the sight of a plough team coming over the brow of a hill, the sight that has been seen in England since England was a land and may be seen in England long

●

after the Empire has perished and every works in England has ceased to function . . . These are the things that make England.[4]

This prophecy proved false. Traditional rural life probably changed more profoundly during Baldwin's premiership than during any previous span of fifteen years. When he began, the thatched cottage where Hardy was born* was the epitome of Wessex; when he ended, speculative housing for the retired was as typical.

Despite the widespread desire to preserve Britain in a blurred picture-postcard haze, the National Trust, then the only organization with the statutory powers to acquire countryside and historic buildings, made little effort to benefit from the property that was changing hands. During the decade before the war the Trust acquired some 5000 acres. The rate of acquisition more than doubled during the war, largely owing to Sir Thomas Dyke Acland's gift, and remained at much the same level until the late 1920s, when once again it more than doubled. Membership was even more sluggish, rising by only 125 between 1914 and 1925. The annual reports at the time urged members to go out and persuade others to subscribe, but in the absence of any publicity drive a few words in annual reports had little effect.

On 28 May 1920, Hardwicke Rawnsley, the last of the three founders, died. At their next meeting the Council expressed their appreciation in exceptionally warm terms, recognizing 'his great enthusiasm for all beautiful things, whether of nature or of art' and 'his indefatigable energy and industry'.[5] *The Times* stated that 'England would be a much duller and less happy and healthy country if he had not lived and worked.'[6] By the time of his death the Trust's estate in the Lake District, of which he had been the moving force, stood at about 5000 acres. Friar's Crag, Lords Island and part of Great Wood on

*Bought by the Trust in 1948 in accordance with his sister, Kate Hardy's, will.

Derwentwater were given in his memory marking a further stage in protecting the scene on which he had looked for thirty-four years. Many of the wooded slopes and islets of this most beautiful of lakes are now in the Trust's care.

Lord Plymouth's death followed two and a half years later and in January 1923 John Bailey was appointed chairman. He had joined the Executive Committee in 1898, and by 1906 he had so impressed Hunter and Rawnsley as to be their choice for vice-chairman. An author of books on Milton and other poets and a regular contributor to literary reviews, he was an improbable man to lead an organization partly devoted to saving the countryside. His friend and future colleague on the Council, G. M. Trevelyan, described him as one of the 'most distinguished literary critics of the first quarter of our century, [who] gave his life up to fitting himself to the critic's task, regarding it less as a profession than as a high calling'. His secondary occupation was 'the defence of the vanishing beauty of the countryside'.[7] He was born in Norwich on 10 January 1864, the son of a solicitor and the grandson of a Norfolk landowner. After school at Haileybury, he read Greats at Oxford, was called to the Bar and stood unsuccessfully for Parliament as a Conservative candidate before settling down to writing. He married the Hon. Sarah Lyttelton, who noted in her introduction to a selection of his letters and diaries which she published after his death that 'few people had more power of enjoyment'. Certainly he was a regular attender of the Literary Society and The Club, two London dining clubs whose members included Asquith, Balfour, Elgar and Archbishop Davidson. He pursued an active social life but his work for the Trust and other charities was carried out only at some financial sacrifice. In 1923 when he was offered the chair of English literature at the East End College of London University he wrote to a friend, 'At first I was tempted partly because it is £1,000 a year of which I am now badly in need (or part of it); but . . . I hate coming down to mere *hack* lecturing [eight hours a week were required]. I want some work which pays a little: most of the work I have done all my life, C.O.S. [Charity Organization Society], National Trust etc. has been entirely unpaid.'[8] Towards the end of his time as chairman, when he was in his sixties, Bailey was described by professor R.S.T. (Theo) Chorley (the father of Roger Chorley, who became chairman of the Trust in 1991), then a young colleague on the Executive Committee, as 'clearly sensitive to natural beauty' and conducting 'the business of the meetings in a competent and sympathetic way if not with outstanding vigour or ability.'[9] However, he had a good mind and he devoted far more time and thought to the Trust than had Lord Plymouth. After the founders, the Trust in its early years owed more to him than to anyone else.

Theo Chorley, who joined the Council in 1927 at the age of thirty-two, and by the following year had initiated the first publicity committee, brought with him the enthusiasm of a new generation and the experience of a different world. He was for many years a professor of law at the London School of Economics, founder-editor of *Modern Law Review* and honorary secretary of the Council for the Preservation of Rural England. He was made a peer by Attlee in 1945 and formed a useful link with the Labour government. With his quick, incisive mind and his ability to see both sides of a question, he contributed much to the Trust during the forty-five years he served on the Council, not least during his period as vice-chairman (1961–9).

Until 1934 the day-to-day business of the Trust was in the hands of S. H. Hamer, who had been appointed secretary in 1911. Hamer was a talented man but not unduly hard-working. Chorley recalls that on a typical day he would:

arrive at the office 11.00–11.30 a.m. reading the letters which had come in, drafting any minutes or memoranda required for the Committee then dictating the results of these activities to his secretary Miss Wilkinson, a charming and capable young woman whose father was a well-known professor at Oxford. He would then walk across the park to his club to take lunch and read the papers, after which at about 3.00 p.m. he would walk back, and sign the letters which in the meanwhile Miss Wilkinson had typed for him and get himself through the day's work by about 3.30.[10]

Despite his easy-going habits, Chorley describes him as a competent secretary who 'carried on the routine of the office, and organized the work of various committees and implemented their decisions in a workmanlike manner for some twenty years. He was of a pleasant and likeable disposition, an altogether likeable man.'[11] Chorley thought that the Trust was being managed 'in a fairly happy-go-lucky way' and doubted whether it was a very go-ahead institution. He notes:

I do not think that much attention was paid to the constitution at this period in the Trust's history: thus in 1926 no less than forty-three members were appointed to the Executive Committee. I imagine that anyone who expressed a desire to help with the work could get himself appointed to the Council. At that time its meeetings were purely formal, and lasted only a few minutes while it listened to a report on what had been done in its name by its Executive. The attendance of the Council members, apart from those who formed the Executive, was small and their interest perfunctory; but occasionally, very occasionally, someone would ask a question.[12]

Bailey's chairmanship falls into two periods. During the first he concentrated on improving the management of existing properties and the selection of those offered rather than on taking energetic steps to enlarge the Trust or its influence. But after 1927 acquisitions, funds and membership all increased and the Trust took a more active part in national affairs. This change in gear was largely due to the arrival of the historian G. M. Trevelyan, as well as Theo Chorley and Oliver Brett, later third Viscount Esher.

Even before becoming chairman, Bailey insisted that the

Trust should exercise stricter standards of selection. He saw the danger of the Trust being looked upon as a means for landowners to hive off unproductive land or old buildings, at the same time ridding themselves of financial liabilities. 'The Trust should exercise a greater discretion in the acceptance of properties and should arrive at a clear understanding of the type of property worth preserving.'[13] One change that was clearly needed was to avoid accepting small isolated properties of no great historic interest or natural beauty and with no control over their surroundings, such as the fragment of Merton Abbey Wall and the two acres of land at Colliers Wood in South London, both accepted before the First World War and now surrounded by roads and a hypermarket.

Bailey improved the administration of the Trust and the management of its finances with the help of R. C. Norman, a brother of the powerful Governor of the Bank of England. Norman, a close friend of Bailey, had been recruited to the Executive Committee in 1918 and soon carried great weight in the Trust. He was chairman of the Finance and General Purposes Committee from 1923 to 1935 and joint vice-chairman from 1924 to 1948. A man of great charm and famous good looks, 'his quick decisive mind grasped the essentials of a problem easily and clearly, [he] acted promptly and without hesitation, and could put his point of view with vigorous eloquence'.[14] In the course of a distinguished career he was chairman of the London County Council and later of the BBC, but he records that 'nothing has given me the same happiness as the National Trust'. His reasons were twofold: 'First the work itself: the preservation from ruin or destruction of areas of natural beauty and of buildings of historic interest and beauty has appealed to me more than anything else I ever did – the buildings even more than the countryside . . . and secondly I never worked with such a sympathetic and congenial lot of colleagues as on the N.T.'[15]

Financial management was improved in two respects. In

1925 control over expenditure was simplified and tightened. From that time the Estates Committee and the Finance and General Purposes Committee submitted annual estimates to the Executive. Once approved the two committees were given a free rein to spend their budget as they saw fit without further reference to the Executive. And in 1927 on the motion of Nigel Bond, then chairman of the Estates Committee, a general endowment fund was created; the intention was to raise £25,000, the income from which would provide for general administrative expenses. This was achieved by 1933.

With the increase in acquisitions the Trust was becoming more aware of the need for endowments for individual properties. At Wicken Fen, Cambridgeshire, for example, the Trust was at first able to derive some income from the sale of sedge for thatching, but as the estate increased in size expenses of administration grew while the income diminished as the demand for thatching material faded away. An appeal was launched in 1926 but it produced little response.

Bailey was anxious to rearrange the management of the Trust's estate. A tiny central office in London relying on enthusiastic and well-intentioned volunteers stretching from Northumberland to Cornwall was not satisfactory when the Trust managed well over 200 properties, as it did by 1932. The Estates Committee, meeting once a month, received regular reports from the active local committees responsible for such properties as Box Hill and the Farne Islands, but little supervision could be exercised over the many smaller properties looked after by a few supporters, or by the donor or his agent. Although little was done at the time, and indeed the Trust did not have the funds to employ more staff, the idea that a regionally based administration would soon be necessary was aired, and in 1932 a northern area representative was appointed to look after the Trust's properties in the Lake District.

The Trust's acquisitions of places of historic interest in the 1920s reflect the contemporary interest in archaeology and in relatively early buildings. The first three decades of the twentieth century saw the restoration of a surprising number of castles and medieval, Tudor and Jacobean manor houses, several of which were later presented to the Trust. Interest in eighteenth-century buildings began to revive in the 1920s and was fostered by the Georgian Group, formed in 1937; Victorian buildings were not in fashion, as Evelyn Waugh noted in *A Handful of Dust*.

The most curious acquisition was Lyveden New Bield in Northamptonshire. This had been brought to the attention of the Executive Committee by two architect members of the Council in 1910 but, despite repeated appeals for funds, the purchase price of £1,000 was not raised until 1922. A roofless shell, it had been begun in 1594 by Sir Thomas Tresham, who spent most of his later life in prison, his punishment for converting to the Catholic faith. Designed in the form of an equal-armed cross, the two-storey building, which was never completed, is a strangely moving monument of piety.

Important archaeological acquisitions included Cissbury Ring on the South Downs, 'the Sheffield of the Flint industry in Neolithic times'[16] and in the Iron Age one of the largest hill forts in England. It has wonderful views in all directions and was bought by public appeal. The Cerne Giant was presented to the Trust in 1920 by two sons of the Victorian archaeologist General Augustus Pitt-Rivers. The figure of this virile giant, cut into the Dorset Downs above the village of Cerne Abbas, stands 180 feet high, the club which he holds in his right hand adding another 35 feet. He is thought to date from the second century AD and may represent the god Hercules. Also dating from the second century is a monument which gives a very different idea of life at the time – Chedworth Roman Villa in Gloucestershire, with its fine mosaics and under-floor heating system. This attracted exceptional interest from the public, and the agreed target was reached in record time. In 1927 most of the much larger sum required for the land round Stonehenge was raised within three weeks, perhaps because on this occasion a letter to *The Times* launching the appeal was signed by the Prime Minister, Stanley Baldwin, and by the Leader of the Opposition, Ramsay MacDonald.

Two years earlier two splendid medieval castles had been left to the Trust by the legendary figure who thought he ought to have been Prime Minister instead of Baldwin, George Nathaniel Curzon, first Marquess Curzon of Kedleston, former Viceroy of India and Foreign Secretary. His will read:

> Convinced that beautiful and ancient buildings which recall the life and customs of the past, are not only a historical document of supreme value, but are a part of the spiritual and aesthetic heritage of the nation, imbuing it with reverence and educating its taste, I bequeath for the benefit of the nation certain properties which I have acquired for the express purpose of preserving historic buildings.[17]

Bodiam Castle in Sussex, one of the most romantic medieval castles in Britain, was built as a defence against the French raiders during the Hundred Years' War. Once opened to the public it became one of the Trust's most popular properties, as it has remained. In 1931, when the Trust had only 2500 members, 10,000 visitors were enough to excite comment, whereas in 1990 with a membership of more than two million, 164,000 visitors were taken for granted.

Tattershall Castle, standing out across the flats of Lincolnshire, had been rescued by Curzon at a time when ancient monuments were still without protection. Curzon describes the events:

> The famous old red-brick castle of Tattershall in Lincolnshire, and the four sculptured stone mantelpieces which were made for it by its founder Lord Cromwell in 1440, and contain the armorial bearings of himself and his family alliances,

and are unique in the world, were sold by the family to whom they had belonged for centuries in 1910. Subsequently, upon the purchaser of the estate becoming bankrupt, they passed into the hands of a Lincoln bank, to whom the Estate had been mortgaged, and were sold by them in 1911. The Castle was presently acquired by an American Syndicate of speculators who looked only to profit. The mantelpieces were sold separately from the Castle and were bought by a London firm of art dealers . . . Finding that there was a very serious and imminent danger that the castle might be pulled down, or otherwise ruined, by the American Syndicate, and learning that there was an interval of twenty-four hours in the course of which it could still be recovered by the payment of a certain profit to them, I intervened to rescue it.[18]

Curzon not only bought and repaired the two castles, he also wrote a history of one and notes on the other as he did for most of the houses he lived in. Despite his busy public life, these were based on meticulous research into state papers, parish records and other sources.

Curzon, a highly intelligent and richly anecdotal grandee, is still remembered in India for his work in restoring on the sub-continent what he thought to be 'by far the most wonderful and varied collection of ancient monuments in the world'. His interest included everything of historical or artistic importance from the earliest civilizations to the Mutiny. After his return to England he served on the Trust's Council as a nominee of the National Gallery and during the years he had to wait before inheriting Kedleston, near Derby, he leased several country houses, including Montacute in Somerset, which he thought the most beautiful house of its size in England (it was later given to the Trust). He devoted much time and money in his later years to restoring and modernizing Kedleston, with the help of the fortune of his first American wife, in the expectation that the house could then remain in the ownership of his family. When this proved impossible his nephew Lord Scarsdale carried out the wish expressed in a supplement to his will, 'I pray that my executors and heirs at Kedleston will see it preserved in good hands,' and in 1987 presented it to the Trust.

As the 1920s went on, the Trust and the public became more and more concerned with threats to the countryside. The increasing number of motor vehicles was bringing in its train strings of petrol stations, advertisement hoardings and specu-lative houses, while the spread of electricity and telephones was accompanied by the march of pylons and telegraph poles. For those who were aware of the Trust's existence a practical response was to give commons or other pieces of land. Most of these were concentrated in Surrey, the Lake District and other counties where the Trust already owned property, but new ground was broken in the Isle of Wight, the Malvern Hills and the Devon coast.

A different type of open space was offered in 1925 by Edward and George Cadbury, grandsons of the founder of the Birmingham chocolate company. They presented 414 acres of their Chadwich Manor estate on the south-west edge of Bir-mingham with the specific proviso that it be maintained as an agricultural estate. It set an example which John Bailey was keen to emphasize when in 1925 he wrote:

> It is a gift of a new kind. What Mr Cadbury and his brother are giving us is not a large open space to be handed over to the public as a recreation ground. We have many of these and hope we shall receive many more. But there is a limit to the amount of land which can be so sterilized. Yet land not sterilized at all, but still devoted to the ordinary purposes of agricultural production, adds immensely to the amenities and the beauty of the near neighbourhood of towns. That is the purpose of this gift, the first of its kind. The object of it is to preserve an agricultural and pastoral oasis in the midst of

what may become a merely urban or suburban district. Access will be given where possible by paths up to the hills. But the farms will remain farms and will not become parks or playgrounds.[19]

Even today public access is only via the footpaths. The Cadburys' generosity has saved this area of countryside from the sprawl of Birmingham and has preserved the views over the Lickey Hills and to the Malvern Hills beyond.

A popular initiative taken by the Trust itself was to launch an appeal for the Farne Islands, which are situated off the Northumberland coast and are one of Europe's most important bird sanctuaries. The appeal had the public support of Baldwin and of Lord Grey of Fallodon, a keen bird-watcher who had also, as Foreign Secretary, watched the lights going out all over Europe in 1914. He wrote to *The Times* early in 1924: 'In one respect they [the islands] are I believe unique. They are the northernmost breeding place of the Sandwich tern and the southernmost breeding place of the Eider duck. Without organized protection the islands would, under modern conditions, be destroyed as a breeding place of the rarer species.'[20] By the time the letter was written, £1800 had already been raised and it was not long before the £2500 needed was subscribed. Not surprisingly the unforgettable sight of thousands of birds nesting among the low plants and on the steep cliffs posed problems for the local society which managed the islands, as it has continued to do for the Trust's wardens. The honorary secretary wrote regularly to S. H. Hamer complaining about requests to visit the islands and insisting that there should be no photography.

A turning point in the Trust's activities came in 1925 with the campaign to buy the Ashridge estate in Hertfordshire. The estate included Berkhamsted Common, which the Commons Preservation Society had saved sixty years earlier, and where at that time G. M. Trevelyan lived. Fearing that it might be sold for residential development, he mounted a national appeal, organizing and addressing public meetings himself. He persuaded one local resident to lead off with a donation of £20,000 and this was followed by a letter in *The Times* signed by Stanley Baldwin, Ramsay MacDonald, Asquith and Grey of Fallodon. The 1600-acre estate was bought after the appeal had raised £80,000, the largest sum yet sought by the Trust.

One of Bailey's greatest services was to persuade G. M. Trevelyan, an old friend of his, to stand for the Council; in 1928 Trevelyan was appointed chairman of the Estates Committee and the same year he became joint vice-chairman of the Executive Committee. He retained these positions until after the Second World War, but his most valuable work for the Trust was as a propagandist. He was also a generous benefactor.

George Macaulay Trevelyan 'was the most famous, the most honoured, the most influential and the most widely read historian of his generation'.[21] He spent his childhood at Wallington, the family home in Northumberland, which his brother later gave to the Trust. He thus grew up a countryman and, as his biographer points out, it was as a countryman that he viewed the world and wrote his histories. He could be caricatured as belonging to the 'cow looking over a gate' school of history (as has been said about Vaughan Williams' music).

He was appalled by the new blights of the motor car and suburban sprawl. In his *History of England*, published in 1926, his description of his country in Anglo–Saxon times echoed Ruskin: 'what a place it must have been, that virgin woodland wilderness . . . still harbouring God's plenty of all manner of beautiful birds and beasts, and still rioting in the vast wealth of trees and flowers – treasures which modern man, careless of his best inheritance, has abolished, and is still abolishing, as fast as new tools and methods of destruction can be invented'.[22] The threat as he saw it was that 'man's power over nature had outstripped his moral and mental development'.

By the 1930s he had become the leading propagandist of

the preservationists, never tiring of reminding others of threats ranging from afforestation to electricity cables, and emphasizing the spiritual dangers. 'Unless he [man] now will be at pains to make rules for the preservation of natural beauty, unless he consciously protects it at the partial expense of some of his other greedy activities, he will cut off his own spiritual supplies, and leave his descendants a helpless prey forever to the base materialism of mean and vulgar sights.'[23]

In 1929 he wrote an eloquent plea on behalf of the Trust, *Must England's Beauty Perish?* After outlining the work and aims of the Trust, which he described as a poor landowner 'with a great number of small estates as widely scattered over the country as the manors of a Norman baron or an Oxford college', he went on to argue that the Trust had become the guardian of the nation's spiritual values. He feared for the 'happiness and soul's health of the whole people. Without vision the people perish, and without natural beauty, the English people will perish in the spiritual sense.'[24] Trevelyan enjoyed his work with the Trust, writing in his brief auto-biography, 'One of my chief interests and pleasures in later life has been to co-operate in the good work with such friends as John Bailey our chief, Hamer the secretary and their success-ors. We were a band of brothers and the work prospered in our hands.'[25] Contemporaries, however, commented on his awk-wardness in committee meetings and his lack of humour, although he was said to possess more than his wife. As a benefactor, Trevelyan concentrated his gifts to the Trust on his adopted county of Cumberland and his home county of North-umberland, which he looked at through the eye of an historian. 'The past is written far more clearly on the face of the borderland than on the landscape of less moorish and more modernized regions.'[26]

In February 1929 the Chesters estate, 21,000 acres of wild Northumberland country, came up for sale; included in it was Housesteads, the best-preserved and most finely situated fort on the Roman Wall. Trevelyan was determined that it should not be closed to the public or carelessly looked after. His father had died the previous year, leaving him a healthy inheritance, and he persuaded the vendor to give the fort to the Trust while he bought the farmland surrounding it, which he then placed under covenant with the Trust. Excavation was carried out that summer. 'What fun it all was! Nothing could be more delightful than to watch the progress of diggings on your own land, conducted by competent and friendly persons. It was a great moment when the spade revealed the secret of third-century tragedy, by exposing a skeleton with a knife in its ribs buried under the floor of a house. Murder will out, even after sixteen centuries!'[27]

After Northumberland it was the Lake District that he loved. For a time he owned a cottage in Langdale, and between 1929 and 1944 he bought land in the district which he presented to the Trust. In 1934 he initiated a scheme for the purchase of an absolutely key 5000-acre estate which included the lakes of Buttermere, Crummock and Loweswater as well as most of their surrounding land. With the help of Balliol College, Oxford and a number of individuals whom he enlisted, the agricultural land was placed under restrictive covenants after being sold on. At the age of fifty-nine he was still able to 'scramble in its inmost recesses. There is a wonderful high lonely tarn called Bleaberry Tarn in the arms of the crags of High Stile where I bathed. I never saw it before. It's a great land.'[28] What taxed Trevelyan most was how to balance the preservation of the Lakes with access for a growing urban population that was becoming increasingly isolated from rural life. Another enthusiast for this magical corner of England was the Rev. H. H. Symonds, a master at Rugby School and later principal of the Liverpool Institute. His devotion was expressed in a variety of ways, through pamphlets attacking conifer afforestation and a guide to walking in the Lakes (a classic in its day) as well as through the gift to the Trust of several farms in the Duddon Valley.

The other major benefactor in the Lake District at this time was Mrs Heelis, better known as Beatrix Potter. As a child she had spent holidays there with her family, several times at Wray Castle, where she came to know Rawnsley. His enthusiasm soon overcame her shyness, he encouraged her painting and twenty years later he advised her on how to publish her first book, *The Tale of Peter Rabbit*. Meanwhile she escaped from London to the Lakes whenever she could, and at the age of thirty-nine she used her literary earnings to buy a small farm, Hill Top at Sawrey above Lake Windermere. Eight years later she married William Heelis, a respected local solicitor, and went to live at Castle Cottage within sight of Hill Top. There she became a working farmer, recognized as an expert on the tough Herdwick sheep of the Lakes.

Through her friendship with Rawnsley, Beatrix Potter had formed a keen interest in the Trust and soon after her marriage she began to subscribe to its appeals. She, like Trevelyan, saw the dangers threatening the Lake District. The problem was and is a delicate one: 'How to preserve the beauty and agricultural integrity of the Lake District so that it should refresh the minds of all who came to it, and be accessible to all; and at the same time to control tourists, parasites, who were so shortsightedly bent on destroying those very beauties and solitudes that attracted their hosts.'[29]

Beatrix Potter shared Rawnsley's view that in the long run the only safe landowner was the Trust. Certainly 'smallholders are hopeless, first they sell off all the sheep stocks; and then they cut down all timber and concentrate on hens'.[30] Despite this criticism she was eager to preserve the small farms so essential to safeguarding the way of life and hence the landscape. But she did not share Rawnsley's desire for the countryside to be enjoyed by everyone: hikers, she thought, should keep to the paths.

By the mid-1920s she was in close touch with Hamer, sometimes writing almost daily with information and advice.

She was clear what the Trust's strategy should be – to buy up farms at the valley heads: 'I can only repeat my heartfelt wish that all the heads of the valleys and roots of the mountains were in the hands of the Trust.'[31] She was also clear that continued grazing was the key to managing the fells: 'They look nothing on the map – alas they will be nothing without the sheep.' And as the tenant farmers could not afford to buy a good stock when they took over, the landlord must do so – a practice that the Trust adopted.

She had little confidence in the financial acumen of the Trust, still less in that of Trevelyan, whom she thought 'not a man of business', who would 'pay any price they ask out of hand'. She was much cannier in making her own purchases, whether the 2000-acre Troutbeck Park, which she bought in 1923, the large Monk Coniston estate seven years later, or the many smaller plots including one at Tilberthwaite that had belonged to her great-grandfather. When she died she left more than 4000 acres to the Trust followed by several farms on the death of her husband: over all this she instructed that 'hunting by otter hounds and harriers shall be prohibited'.[32]

It was an open secret that the land she owned would be left to the Trust. However, during her lifetime she was adamant that she would look after it herself, taking the daily decisions of where to plant trees or introduce new flocks of her beloved Herdwick sheep. This did not prevent her from writing a barrage of long and frank letters about the Trust's properties, tourism, motorcycles, fires and personalities. Of Bruce Thompson, the admirable representative in the Lake District, she wrote to Hamer: 'Although it is probably useless, I am writing about your agent Mr Bruce Thompson . . . A man clearly cannot help having been born dull. Thompson is supercilious as well.' On another occasion she wrote, 'I must point out that I did not say I felt Mr Thompson to be "unsatisfactory as the Trust's representative in the Lake District". What I said was that he shows no judgment in dealing with trees and woods at Coniston.' Clearly success had transformed a once timid character: as she told a friend, 'I begin to assert myself at seventy.'[33]

From the outbreak of the First World War until the arrival of Trevelyan and Chorley the Trust undertook little work of a 'militant' character. The vacuum thus left was filled in 1926 when the Council for the Preservation of Rural England was formed. This was a federation of twenty local organizations and forty-one affiliated societies. Its first honorary secretary and prime mover was Patrick Abercrombie, the eminent town planner and author of the idea of Green Belts round London and other cities. The twenty-seventh Earl of Crawford, previously a minister in the Lloyd George coalition, who had been a member of the Trust's Council since 1900, a founder of the National Art-Collections Fund and chairman of the Royal Fine Art Commission, was elected the first president. The aims of the Council were threefold: to educate public opinion, to organize concerted action and to promote relevant legislation. It was financed by a few benefactors and immediately started pressing the government to introduce planning legislation throughout the country. 'Blatant advertisements, ugly gasometers and the country's quiet rural charm destroyed by people who commit bungalows' were seen as the most immediate threats to the rural scene.

Lloyd George, Stanley Baldwin and Ramsay MacDonald all welcomed the arrival of the Council, while the *Daily Telegraph* speculated, 'What will it [rural England] be in the second half of the century one is almost afraid to think, unless a new conscience is developed among us, a new sensitiveness to things of beauty and a resolute determination that the face of England shall not be wantonly scarred and disfigured.'[34]

The CPRE had a powerful polemicist in Clough Williams-Ellis, the architect and creator of the Italianate village of Portmeirion on the coast of North Wales. *England and the Octopus*, published in 1928, warned of the monster of the materialist age whose tentacles were slowly suffocating the countryside. 'Everyone who reads this book, indeed everyone who reads at all or has eyes in his head knows that England has changed violently and enormously within the past few decades. Since the War, indeed, it has been changing with an acceleration that is catastrophic, thoroughly frightening the thoughtful among us and making them sadly wonder whether anything recognizable of our lovely England will be left for our children's children.'[35] The book expressed feelings shared by many that England was in danger of becoming 'one industrial city of philistines dotted with occasional country oases'. But Williams-Ellis warned against taking preservation too far instead of planning a more orderly and beautiful land, where amenities and the visual arts would have their due place. 'It is really one of our morbid symptoms that if any place of natural beauty or historic interest is to be sold, we forthwith attempt to raise a fund for its purchase. It is called "saving it" – why? Because we know well that if it is bought it will almost certainly be built over, desecrated and destroyed. We have indeed almost become a museum in which are preserved here and there carefully selected and ticketed specimens of what England *was*. The National Trust is England's executor.' The final chapter of this influential book was written by Patrick Abercrombie, who proclaimed his belief in the CPRE and acknowledged the contribution of the National Trust.

By 1929, when the second Labour government took office, the Trust had revived its lobbying activities, putting down amendments to Bills and giving evidence to government committees. Acting with the CPRE, it was sufficiently influential to persuade Ramsay MacDonald to appoint a committee chaired by Christopher Addison, then Parliamentary Secretary to the Minister of Agriculture, to look into the possibility of National Parks. Addison's brief was to 'consider the desirability and feasibility of such parks for preserving the natural landscape and wildlife and improving facilities for recreation'. The Committee recommended that National Parks should be established, though it was not until the Attlee government that the legislation was passed. The Committee went on to recommend that the Parks should be administered by the Trust. This proposal, though flattering, was turned down by the Executive Committee on the grounds that the Trust must remain an unofficial body and that only as such could it effectively carry out its aims. The principle that the Trust must retain full independence without even quasi-official functions has since been regarded as essential to its authority and standing.

The CPRE was followed by other new groups. In 1933 came the Youth Hostels Association, with Trevelyan as its president, to provide cheap accommodation for walkers and cyclists; and in 1935 the Ramblers' Association was set up, an amalgamation of groups determined to gain access to private land. As these groups gained in members and influence the Trust gradually left campaigning to them, intervening occasionally on major issues or where its own properties were involved.

Walking or rambling clubs first appeared towards the end of the nineteenth century, with such names as Sunday Tramps, until in 1905 a federation of rambling clubs round London was set up at a meeting called by Lawrence Chubb, the secretary of the CPS, and chaired by Sir Robert Hunter. The aims then were not very different from those today: to campaign for the right of ramblers to be as free as grouse to roam the mountains and moorlands of England, while better protected from the dangers of being shot. During the 1920s scores of local rambling federations were born, and in 1928 a national council was formed which in 1935 became the Ramblers' Association. The Trust forged close links with this as well as with the CPRE and the YHA, and in due course all three were given the right to nominate members to the Trust's Council.

In Scotland the CPRE's counterpart, the Association for the

Preservation of Rural Scotland (APRS), later renamed the Association for the Protection of Rural Scotland, was the catalyst in creating a separate National Trust north of the border. Despite campaigning on Scottish issues since 1895, and despite attempts by Rawnsley to establish a local branch in 1908 and 1914, no properties in Scotland had been presented to the Trust, and after 1920 there was not even a local corresponding member in Scotland.

The critical event came in 1929 when the APRS was offered an estate in Galloway which it was unable to accept because it did not have the power to hold land. The property could have been offered to the National Trust but there was a strong feeling in the Council of the APRS that Scotland should have its own Trust. On being consulted, the National Trust fully concurred with this view as Hamer made clear when he was invited to speak at the first meeting of the Scottish Provisional Council on 10 November 1930: 'We also realize that the conditions here in Scotland are very different from what they are in England, and it would be quite impossible for anything in the nature of a National Trust to be conducted from London.'*

In notes circulated to the Provisional Council, the secretary of the APRS, Frank Mears, outlined what he saw as the essential difference compared with the position in the more densely populated country of England: 'Our problem in Scotland is rather that of the preservation of glens or valleys or groups of hills as a whole, and thus we come up against important social and economic factors.'[36] He doubted whether a Scottish Trust should accept specific areas of limited size, believing that it should 'rather evolve some means of preservation not only of beauty but of the old life which is part

---

*The events leading up to the formation of the National Trust for Scotland are described in *The Formative Years 1929–1939*, a booklet by George S. Russell, who succeeded his father as the Scottish Trust's law agent in 1951; together they served in that capacity for fifty-one years.

of it. To preserve in fact the territorial relationships which mean so much.' This nostalgic ideal may have been unattainable, but the needs of local communities were from the start a potent concern.

The National Trust for Scotland was formally constituted in 1931. Close links with the National Trust were established from the start through the Earl of Crawford, who had been an early member of the English Council and was one of the four signatories of the initial letter circulated to potential supporters in Scotland. (His son later became a vice-president and also chairman of the English Trust.) The other signatories were also substantial landowners – the Earl of Haddington, Sir Iain Colquhoun of Luss and Sir John Stirling Maxwell of Pollock, a very different group from the founders of the National Trust. But there were similarities. Scotland had the eighth Duke of Atholl as first president to rival the first Duke of Westminster and both relied on the services of a devoted lawyer – in Scotland Arthur Russell, an Edinburgh lawyer in the mould of Robert Hunter.

The Scottish Trust quickly attracted support and by the outbreak of war eight years later it owned twenty-eight properties. Among early ones were battlefields resonant of Scottish history which would never have come to a London-based organization: part of the field of the Battle of Bannockburn; Bruce's Stone in Galloway where Robert the Bruce defeated the English; the Signal Rock near the site of the Massacre of Glencoe; and the Glenfinnan Monument which was erected in 1815 by Alexander MacDonald of Glenaladale in tribute to the clansmen who fought and died in the cause of Prince Charles Edward Stuart. Part of the battlefield of Culloden was bought more recently. The sense of national pride also inspired the gift of homes of several of Scotland's best-known writers and some of Scotland's most spectacular scenery. The first mountainous estate was Glencoe. Arthur Russell, the Trust's law agent, had been instructed to proceed to Glencoe and bid for the Signal Rock. But the night before the auction he found a rival purchaser who was prepared to buy part of the glen, including the reputed site of the Massacre. This attracted the interest of his climbing friend Percy Unna, president of the Scottish Mountaineering Club, who became the 'anonymous donor' of much of Glencoe, Dalness and other extensive mountainous areas.

The Scottish and English Trusts have many features in common – their historic sites, houses (acquired in many cases under the Country-House Scheme, as was the House of the Binns), and gardens and their mountains, moors and forests. But they have pursued rather different courses which reflect the differences in size, topography and economy of the two countries. The smaller population has made it easier as well as more necessary for the Scottish Trust to work in partnership with government, local authorities and other organizations.

The second property bought by the Trust in 1932 was Culross Palace. This was made possible only by the initiative and courage of the Trust's founders in using half of the Trust's first legacy towards the purchase, and it underlined the Trust's concern for the simple vernacular architecture of Scotland. The Palace was accepted by the government for repair and management (and handed back to the Trust in 1991) and this set a precedent for joint ventures later on. The Scottish Trust continued, in the absence until fairly recently of the influential preservation societies existing south of the border, to give a lead in campaigning – by encouraging the listing of historic buildings, defending beautiful landscapes from hydro-electric schemes and becoming involved in many major preservation matters in Scotland. The first lists of historic houses were introduced by the fourth Marquess of Bute as an example to the government and became known as the 'Bute Lists' or the 'National Trust for Scotland Lists'.

The two trusts have set different priorities for preservation. In Scotland the mountains have had more appeal than the coast: in the early 1960s when the English Trust was preparing for Enterprise Neptune, the Scottish Trust commissioned a survey of the Highlands to identify areas of outstanding natural beauty and focus public attention on the need to protect them. The Scottish Trust's most striking contribution to coastal protection has been the acceptance by gift or bequest of some of the smaller islands – Fair Isle, St Kilda, Iona, Canna and others. For buildings, priorities have also been different. Scotland has fewer great houses than England but a wealth of small old houses, many of them having become ruinous by the mid-twentieth century. One of the Scottish Trust's first projects was to buy and restore derelict properties in the Royal Burgh of Culross and thus help to bring life back to the town. In 1960 this initiative was followed by the Little Houses Improvement Scheme, which with help from the Pilgrim Trust has played a major role in the revival of the Fife fishing ports and other derelict settlements and has been successfully copied in Northern Ireland.

The two Trusts work closely together, each being represented on the other's Council and their members having the right to visit each other's properties free of charge. Scotland has benefited from England's wider experience and more specialized expertise, and England from Scotland's flair and international outlook. It was the Scottish Trust which initiated the first International Conference of National Trusts and held the opening meeting at Culzean Castle in Ayrshire. The Scottish Trust also organized the second conference, which with the help of their American supporters, Scottish Heritage USA, took place in the eastern seaboard of the United States.

Another contentious issue that troubled the National Trust and the CPRE was the threat to the landscape of large-scale conifer plantations promoted by the Forestry Commission. Not surprisingly the most vocal opposition came from Trevelyan, who feared that areas such as the Lake District were in danger of resembling some 'German Pine Forest'. Following the felling of woodland during the First World War Britain was left with only 3 per cent of its land under forest, the lowest proportion in Europe. The Forestry Commission's brief was to 'improve the supply, production and reserves of timber both by its own operations and by promoting the idea of forestry to private interests'. A target of doubling the land afforested was set and the Commission was given power to make grants for planting.

By 1929 138,000 acres had been planted, and by 1939 this figure had risen to 425,000 acres. Such rapid expansion coupled with a determination to become profitable led to poor forestry techniques and large areas being blanketed in conifers. At this time the concern of the Trust and the CPRE was not any fear that wildlife and their habitats were endangered but that the public would be denied access and the beauty of the landscape would be destroyed. In response to criticism the Forestry Commission set up a committee which included representatives from the CPRE and the Trust. The result was encouraging and in 1936 the Forestry Commission began to provide access for the public by opening its own Forest Park in Argyllshire shortly followed by other parks in Snowdonia and the Forest of Dean.

Trevelyan, however, was not satisfied that the Lake District was free from trouble. The government had issued instructions for large-scale tree planting in order to help unemployed miners from the coastal area, and the proposed land included the central part of the Lake District. Trevelyan objected, pointing out that Wordsworth had 'long ago denounced the introduction of the conifer as a crime against Nature's local by-laws' and such planting would threaten to wipe out 'the old English forest trees – oak, beech, ash, elm, and sycamore'.[37] As chairman of the Estates Committee, he privately negotiated with Sir Roy Robinson, the Forestry Commission chairman, to prevent any planting in the holiday areas of Borrowdale, Langdale, Rydal, Grasmere and Ullswater.

His success in the Lake District did not stop the question of forestry being a continuing worry. At a meeting of the Trust's

Local Committee Conference in 1949, William Vane, a Conservative MP and Cumbrian landowner (later the first Lord Inglewood) and at that time a new member of the Estates Committee, announced that he thought that the lingering controversy between the Forestry Commission and the Trust was 'just too childish'. At this point Trevelyan roared from behind him, 'Just what do you mean by that, young man? You must know that the National Trust has played a great part in preventing the planting of conifers in the Lake District.'[38]

The Trust benefited enormously from the surge in national interest in the countryside. Nineteen-twenty-nine, described by one Council member as a year of 'wonderful weather and wonderful gifts', went down as an *annus mirabilis*. The Trust's membership had at last reached 1000; it owned almost 200 separate properties and 37,000 acres; and, although its subscription income amounted to only £1720, it received £14,899 in legacies and donations and its other income amounted to £30,768. Stanley Baldwin and Ramsay MacDonald had become honorary vice-presidents, and the work of the Trust was gaining momentum despite the October Stock Exchange crash, presage of a world slump.

The Trust was becoming a national institution, looked to for guidance and assistance when questions of preserving places of historic interest or natural beauty arose. But it had not been successful in securing any important changes in legislation since before the First World War. The opportunity came with the Labour government of 1929–31. Several measures which the Trust had some share in bringing about were introduced, including a strengthened Ancient Monuments Act, a Town and Country Planning Act extending planning from urban areas to the whole country and, more important for the Trust, provision in the Finance Act of 1931 to exempt from death duties property given or devised to the Trust.

The annual report for 1930–1 notes the marked growth of public opinion concerned with preservation. This was

expressed in a letter from the Prime Minister, Ramsay Mac-
Donald, to the Trust:

> The Englishman may say little, but down in his heart there
> burns an intense love of his country and pride in his heritage
> – an unique civilization and a history every surviving monu-
> ment of which is precious, a countryside profuse in natural
> beauty and unrivalled in the variety of its scenery . . . To the
> energy and vision of the National Trust we owe an immense
> debt of gratitude. Their reports showing the growing list of
> benefactors and telling of new acquisitions in all parts of the
> country make cheerful reading indeed – Duty requires that
> much more should be done for the preservation of our heri-
> tage.[39]

It is notable that even a Labour Prime Minister was content to
rely on voluntary effort rather than turning to state or
municipal authorities, as was happening on the continent and
in North America.

The spate of publicity had one unexpected consequence.
On 30 December 1930, £5 was presented to the Trust by an
unknown secret club calling itself Ferguson's Gang. A group of
eight well-educated young women in their twenties had begun
to collect Victorian coins. Unsure how to spend their
accumulated wealth one member who had read Williams-Ellis'
England and the Octopus suggested that in order to frustrate the
monster referred to in the book their funds should in future be
presented to the Trust. A jingle written by one of the Gang
encapsulated their feelings: 'The glorious day isn't far away,
When London and Liverpool meet, And nothing remains of
England where the country used to be, But roads run straight
through a housing estate, And a single specimen tree.'[40] In
1931 the enigmatic Gang bought their first property, Shalford
Mill, an eighteenth-century watermill in Surrey that later
became their headquarters, and was repaired with the help of
an architect friend known as the Artichoke. Most of the Gang

had not much money but one member could be relied upon to supplement their funds. Their other gifts included the old Town Hall at Newtown in the Isle of Wight, the medieval remains of Steventon Priory, then in Berkshire, and a stretch of the Cornish coast. In 1935 the brother of one of the Gang (masked to conceal his identity) walked into Broadcasting House and read out an appeal which raised £900 and brought in 1000 new members, increasing the total by 25 per cent at a stroke.

The members prided themselves on their anonymity, using such names as Bill Stickers, the Bloody Bishop and Sister Agatha. 'Ferguson' was a mythical character, chosen for his unremarkable but solid name. At intervals a masked member would walk into the Trust's office and deposit a sack of cash, once as much as £2000, on the secretary's desk. The whole enterprise was as beneficial to the Trust as it was original in concept and efficient in execution. Although reports of their exploits appeared in the popular press, they remain a mystery. It is known that several are still alive, despite the clause in their constitution that states, 'The Membership of Active Members is terminable only by death; but this is not so difficult as might be supposed.'[41]

The Trust's growing reputation also attracted more conventional benefactors. Among then was Noel Buxton, a member of the well-known Norfolk family, who later amalgamated his names, becoming a Labour peer and MacDonald's Minister of Agriculture. He bought and repaired a fine half-timbered house in the village of Coggeshall, Essex which had formerly belonged to his family and passed it on to the Trust for preservation. Another benefactor was the antiquarian Aymer Vallance, who gave another half-timbered village house, Stoneacre in Kent, with the furniture collected by himself and his wife. He was obviously aware that by this time the Trust was becoming more discriminating. 'It is a great relief to both of us to be assured that the house on which we have expended so much care and labour and money, will be kept in safe hands in perpetuity.'[42]

In 1931 came the first of two properties given by one of the Trust's most generous benefactors, Ernest Cook, grandson of the nineteenth-century travel entrepreneur: Montacute, the Elizabethan house in Somerset, earlier leased by Lord Curzon, and the Bath Assembly Rooms. This latter building was the first eighteenth-century setpiece to be presented to the Trust, its austere classical façade concealing three magnificent rooms within. The building, like others of its period, had suffered neglect and was in poor condition. The Trust would not have been able to accept the Rooms had not the City of Bath agreed to carry out repairs and take a long lease. Cook, who wished to remain anonymous, conveyed his gifts through the SPAB. The Trust had come to an arrangement that it would hold buildings on behalf of the SPAB on the understanding that the latter would be primarily responsible for the maintenance.

A new and continuing source of finance for the Trust came into being in 1930. This was the Pilgrim Trust, a charitable foundation established by the American philanthropist Edward Harkness, whose father had made a fortune from oil and railways. Stanley Baldwin, the first chairman, and his co-trustees decided to devote a substantial proportion of their resources to the preservation of historic architecture, works of art, archives and the beauty of the countryside, in accordance with their own beliefs and with what they thought would have been Harkness's wishes. One of their early grants was offered to meet the administrative costs of the Scottish Trust for its first three years and another went to the National Trust for general purposes. By 1939 the Pilgrim Trust had given the Trust more than £28,000, mostly in response to specific appeals. This generosity had much to do with Trevelyan's friendship with the chairman, Stanley Baldwin, and with another trustee, John Buchan, novelist and Governor-General of Canada.

John Bailey died in 1931, leaving a Trust barely recognizable from the one which he had inherited. The support of two prime ministers and the new emphasis on publicity had brought the Trust to the notice of a wider public. And this was accompanied by the acquisition of properties symbolic of English history – the site of Magna Carta at Runnymede, the land surrounding Stonehenge, and part of the Seven Sisters, the great cliffs sweeping up to Beachy Head near Eastbourne. Trevelyan declared that the Trust had become 'an ark of refuge and a bulwark in the day of trouble'.[43]

CHAPTER FOUR

The Country-House Scheme

JOHN BAILEY's successor, the second Marquess of Zetland, took over as the Great Slump was drawing to a close and remained in office until near the end of the Second World War. Nonetheless he presided over the most significant new direction the Trust had yet taken – the preservation of historic houses – and over a rapid acceleration in land acquisition. The two developments were not unconnected for some of the historic houses were endowed with extensive country estates.

'It is a fact,' wrote Clough Williams-Ellis in 1928, 'patent to all and deplored by some, that the large-scale private paradise is already obsolescent. There are even now more great country houses in England than there are rich men able and willing to inhabit them.'[1] But the few preservation bodies that existed, including the Trust, showed little interest in the matter, nor did the Ancient Monuments department of the Office of Works. Even the country-house owners were mostly not aware that the *ancien régime* was coming to an end until they were beset by the privations and financial difficulties of the war and its aftermath.

In 1930, two years after *England and the Octopus* had been published, the magazine *Country Life*, which considered itself to be the voice of the countryside, took on a new and positive tone, pressing the government to:

> do something practical to prevent the constant breaking up
> of beautiful properties into ugly building estates and the
> dispersal abroad of well-nigh priceless collections. The state
> has already exempted from death duties works of art of

public interest. Why should it not extend this principle to parks, woods and open spaces such as those of Goodwood, which, though privately owned, are always open for public enjoyment, and to houses such as Knole, which are in reality treasure houses of beautiful things?[2]

Influenced by Christopher Hussey, who was the dominating personality on the magazine from the late 1920s and who later left Scotney Castle in Kent to the Trust, this was the first of a series of leaders and articles about country houses under threat.

The underlying reasons for the decay of the country-house economy were already evident by the end of the nineteenth century – the decline in agricultural income and the introduction of death duties coinciding with the extension of the franchise and the erosion of the landowners' political influence.

The break-up of the landed estates began slowly, but gathered pace in the years immediately before and after the First World War. In his massive study, *The Decline and Fall of the British Aristocracy*, David Cannadine concludes that the British grandees and gentry, who in the 1870s had been the wealthiest in Europe, enjoying a high social position and an unusual political dominance, both at the centre and in the locality, had by 1939 lost their relative wealth, status and power. The change was recognized as early as 1934 by Winston Churchill when he gave the address at the funeral of his cousin, the ninth Duke of Marlborough:

> During the last forty-two years the organism of English society has undergone a complete revolution. The three or four hundred families which had for three or four hundred years guided the fortunes of the nation from a small struggling community to the leadership of a vast and still unconquered Empire lost their authority and control. They became merged peacefully, insensibly, without bloodshed or strife, in a much more powerful but much less coherent form of national consciousness. The class to which the late Duke belonged were

not only almost entirely relieved of their political responsibilities but they were to a very large extent stripped of their
property and in many cases driven from their homes.[3]

As the number of large houses being pulled down or having
their estates broken up increased in number after the First
World War, more people began to voice their concern, and in
1933 Professor W. G. Constable, director of the Courtauld
Institute, challenged the National Trust to take on the responsibility of saving the country house. The leading figures in the
Trust were, of course, not unaware of the problem, but they had
so far taken no public stance. When John Bailey died, R. C.
Norman, who had been acting as chairman during Bailey's last
illness, was asked to succeed him. He refused. He felt that at a
time when the Trust depended largely on the goodwill and
favour of the great landlords, it was important that its chairman
should be one of their own whose words they would heed and
whose example they would follow. His eye fixed on Lord
Zetland (a cousin of his wife), whose previous public service
had lain mostly in India. Zetland accepted and proved to be a
good choice.

Lawrence Dundas, second Marquess of Zetland, was born
in June 1876. After Harrow and Trinity College, Cambridge, he
spent much of the next few years on solitary travels in India and
the more remote areas of Asia, interrupted by a short spell on
Lord Curzon's staff. From 1907 until 1916, when he was
appointed Governor of Bengal, he was Conservative Member of
Parliament for Hornsey. After some years of relative leisure,
during which under the intermediate title of Ronaldshay he
wrote books on India as well as the official biography of Curzon,
he was in 1935 appointed Secretary of State for India, a position
he held until the formation of the coalition government in 1940.
He was the second of three chairmen in the fifty-three years
between 1914 and 1965 to have served as a Conservative minister or MP – the fourth had tried unsuccessfully to obtain a seat.

Zetland realized that the Trust was facing an era of concentrated effort and expansion, and he devoted considerable attention to its affairs even after joining the government. He used to invite Macleod Matheson, who had become secretary of the Trust in 1934, to the India Office to go over the agenda for the Executive Committee so that he had all the material aspects of the subjects under discussion at his fingertips and could evaluate the interventions by the different speakers.

Chorley notes in his memoirs:

As chairman of a committee Zetland can have had few equals in his time, certainly he was the best I ever served under. It was not until one had seen him at work that one began to appreciate his true worth. He was a slightly dandified figure who looked rather as if he wore a corset; his voice was rather unattractive and his delivery somewhat pompous. Indeed, as a speaker he had little ability either to make his subject attractive or to hold the attention of his audience, all of which disadvantages were underlined by an irritating gesture he made by a jerking movement of the head repeated continually.

Despite these faults Chorley insists that the work in committee was never so well done as when Zetland presided. Chorley found him 'less likeable than his predecessor, John Bailey, and much less so than his successor, Lord Crawford, both of whom had a more imaginative grasp, greater sensitivity and knowledge of the subject matter with which the work of the Trust is concerned.'4 James Lees-Milne, the Trust's country-house secretary, describes him as wry, starchy, pedagogic and humourless: 'The memory of Lord Zetland standing on the pavement outside the office, wearing a neat trilby set squarely over a poker face, turning neither to the left nor to the right, but brandishing aloft a furled umbrella as though it were the sceptre on offer to his sovereign in Westminster Abbey and intoning the long-drawn-out word, "Cab!" still reduces me to a nervous tremor.'5

It was Zetland's idea, which proved inspired, to invite Lord Lothian to speak at the Trust's 1934 annual general meeting. On 19 July, in the Inner Temple Hall, Lothian delivered a stirring speech which changed the course of the Trust's history and marked the beginning of country-house preservation in the UK. Lord Lothian was well qualified to give a speech of this nature. After an early career as a member of Milner's 'kindergarten' and Lloyd George's 'garden suburb', Philip Kerr, a bachelor, had succeeded his cousin as eleventh Marquess of Lothian in 1930. With the title came four country houses and crippling death duties, which forced him to sell, among other things, a number of greatly prized rare books. Replying to a letter from Lloyd George congratulating him on his elevation he wrote, 'I shall have to pay to our exhausted Exchequer almost 40 per cent of the capital value of a mainly agricultural estate. In my capacity as an ordinary citizen I think highly of these arrangements but as an inheritor of a title and estates thereto they will prove somewhat embarrassing.'[6] He dealt with this embarrassment and with a bill from the Exchequer for £300,000 by divesting himself of three of the four houses: Ferniehirst was let to the Scottish Youth Hostels Association, Newbattle became an educational centre for the Scottish Universities and Blickling in Norfolk was left to the Trust. When Lothian died in December 1940 he was Ambassador in Washington, where he was proving remarkably successful in mobilizing American support for the British war effort. (His death at that time was avoidable, for as a convert from Catholicism to Christian Science he would see no doctor.) He had spent his life in public affairs. Apart from his six years with Lloyd George, he had served briefly during 1931–2 as a Liberal member of MacDonald's national government.

Lothian's speech, entitled 'England's Country Houses – the Case for their Preservation', set out the arguments as cogently as has ever been done:

I venture to think that the country houses of Britain, with their gardens, their parks, their pictures, their furniture and their peculiar architectural charm, represent a treasure of quiet beauty which is not only specially characteristic but quite unrivalled in any other land. In Europe there are many magnificent castles and imposing palaces. But nowhere, I think, are there so many or such beautiful country manor houses and gardens, and nowhere, I think, have such houses played so profound a part in moulding the national character and life. Yet most of these are now under sentence of death, and the axe which is destroying them is taxation, and especially that form of taxation known as death duties.[7]

He went on to describe the increasing burden. Under the Finance Act of 1904, estate duty stood at 8 per cent. After the Budget of 1909 and until 1914 the maximum rate was 15 per cent; in 1919 it rose to 40 per cent and then in 1930 to 50 per cent. This rate, along with the continued depression in agriculture, led Lothian to conclude, 'Looking at the picture as a whole I do not think it is an exaggeration to say that within a generation hardly any of the larger historic houses of Britain . . . will be lived in by the families who created them.'

Lord Lothian proposed that the Trust or possibly a government commission should survey the houses that were of real historic interest and artistic merit. The criteria were to be fourfold: 'First, that the house or the major part of it should be at least 100 years old. Second, that it should be of definite historic or architectural merit. Third, though this does not apply in all cases, that it should have a garden or park around it. Fourth, that it should be suitably furnished and maintained as a dwelling-house.' Lothian urged that the houses should, if possible, continue to be lived in by their owners and, if not, by new tenants who were prepared to respect and preserve their surroundings. 'Nothing is more melancholy', he commented, 'than to visit these ancient houses after they have been turned into public museums, swept, garnished, dead, lifeless shells, containing no children's voices, none of the hopes and sorrows of family life, no procession of guests meeting to discuss politics or economics, literature or art, or the varied problems of contemporary life.' The difficulty was that, while certain contents of a house were exempt from death duties, the house and gardens – although just as valuable a part of the national treasure – were not. Lothian argued that, in return for some measure of public access, it was folly not to grant exemption to the house, gardens and historic contents, provided that they were preserved as a whole. While such measures would help, Lothian recognized that some houses would not continue to be lived in by their owners. However, he thought that a considerable number of these might be given or left to the public if their owners felt that they would thereby be preserved. In conclusion he set a challenge for the Trust: 'I believe that if a body like the National Trust were willing to equip itself to become a landlord on an ampler scale, it would gradually draw within its ambit quite a large number of historic furnished houses.'

For a time there was silence while the Trust thought hard about its future. It would not be easy to obtain relief from the burden of taxation in the interest of preserving country houses. Why should the government award tax concessions to a group of wealthy families in return for limited public access? Behind the scenes, however, the negotiations that led to the National Trust Act of 1937 were already under way, through Zetland at a political level and through Matheson with officials.

Macleod Matheson, a man of nervous intelligence with sandy hair and a beard, was secretary of the Trust between 1932 and 1945. He was educated at St George's School, Harpenden, and Balliol College, Oxford, after which he served in France during the First World War. In 1922 he became assistant secretary and later secretary of the Gas, Light and Coke Company. But his health was not good, the strain became too great and he left his post. However, still only thirty-six years old, he felt strong enough to take on the secretaryship of the National Trust, then a midget compared to the might of the Gas, Light and Coke Company. It is likely that both the Trust and Matheson made a miscalculation about the work that was involved. Matheson had accepted the post believing that it would be far less demanding than business. Chorley remembers that Matheson was asked at his interview whether a large salary drop would matter. 'He hesitated and said that after talking with his wife they were both of the opinion that his health would stand up much better to the work at the Trust. No one had envisaged that the work of the Trust would expand so fast.'[8]

Zetland recorded that Matheson rendered yeoman service. He undoubtedly conducted with sensitivity the complicated negotiations involved in promoting the National Trust Act of 1937 and also the discussions with powerful families such as the Trevelyans about taking on their estates. But Chorley's notes on Matheson are not complimentary: 'He lacked drive and vitality. He always seemed to be working at a low ebb.' Nor was he popular with Lees-Milne and the other staff: 'He was a jealous boss and a bad delegator of business. Needless to say, he excelled over the rest of us in grasping the most recondite problems, which was always annoying.'[9]

One of the moves made by the Trust following Lord Lothian's speech was to contact *La Demeure Historique*, a French forerunner of the Historic Houses Association. This elite trade union of owners had come into existence in 1923 in an attempt to persuade the French government to tackle '*la grande misère des châteaux de France*' and to reform laws governing taxation and rating. They had also taken steps to raise funds themselves by organizing sight-seeing tours of châteaux within each *département*, which produced not only income for the owners, but also a central fund from which occasional grants could be made. In 1936 the Duc de Noailles, founder of *La Demeure Historique*, was invited to address a selection of English country-house owners at the Royal Geographical Society's rooms in Kensington. His words were a factor in the Trust's initial perception of its role as more adviser than owner. But Lothian's call had already struck a chord and the annual report for 1935–6 was bold enough to suggest that 'Probably a number of these houses with their parks and gardens and contents will one day be held by the Trust for the benefit of the nation. . . .'[10]

The French government had given little help to the owners of châteaux beyond making a small tax concession on income from visitors' fees and urging local authorities to have regard to the expense of maintaining old buildings when fixing the rates. But at least they had recognized that there was a problem and had promised that at some unspecified date in the future the grants already available for a handful of the most outstanding châteaux would be offered to those listed on a supplementary register. The British government recognized no such problem: Zetland had to report that they had shown no real interest in the plight of the owners and had generally given a cold shoulder to the idea of any tax concessions.

Their response reflected the general view of what should be the responsibilities of the state. This was summed up by Maynard Keynes, eminent economist and great patron of the arts, in an essay he contributed to a volume edited by Clough Williams-Ellis in 1937, entitled *Britain and the Beast*. He cited as a particularly good example of the way Britain was hag-ridden by what he considered to be a perverted theory of the state the government's refusal to take action on the preservation of the countryside, when the obvious remedy was to prohibit its exploitation and pay just compensation. Another example he cited was the preservation of national monuments, which was regarded 'as properly dependent on precarious and insufficient donations from individuals more public-spirited than the community itself'.[11] In Keynes' view both examples stemmed from the misguided belief that it was positively wicked for the state to spend a halfpenny on uneconomic purposes.

The British government's response to Zetland's initial approach was to ask the Trust to draw up a list of houses of 'undoubted merit' and a year later not to obstruct the Private Bill promoted by the Trust. At this stage the Trust's Executive Committee appointed a sub-committee – the Country Houses Committee – to work out detailed proposals. At its meeting on 12 March 1936, Lord Zetland took the chair for the first and only time. Sitting with him were Lord Esher, the future chairman of the committee; Sir Alexander Lawrence, a former solicitor to the Treasury; the fourth Baron Methuen, owner of Corsham Court, Wiltshire; Nigel Bond, a former secretary of the Trust; R. C. Norman and Major Michael Peto, another country-house owner. Also in the room were Macleod Matheson, the Trust's secretary, and James Lees-Milne, who had been appointed secretary to the new Committee. Responsibility for selecting the houses of merit was delegated to a small group consisting of Christopher Hussey, editor of *Country Life*; Sir Charles Peers, retired Chief Inspector of Ancient Monuments; Lord Gerald Wellesley, later seventh Duke of Wellington, architect and one of the original members of the newly founded Georgian Group; and W. A. Forsyth, another architect and strict supporter of the SPAB's 'anti-scrape' doctrine.

Within three weeks the group had drawn up a list of 230 houses, including not only Chatsworth, Blenheim and other palaces, but also some smaller houses such as Melford Hall and Cotehele which later came to the Trust. Although too new and too diffident to say anything at the time, Lees-Milne felt that not enough thought was given to this first list and hundreds of houses had been omitted, including some that were not fashionable at the time but that later generations would find of merit. Nevertheless, the list was approved at the next meeting of the Committee, which had added Sir Edgar Bonham-Carter and the fourth Marquess of Salisbury to its number. Like the Estates Committee, the Finance and General Purposes Committee and the Publicity Committee, the

Country Houses Committee was nominally only advisory to the Executive Committee. But Lees-Milne recalled that 'so elastic was the Trust's constitution in those heady days that the Country Houses Committee never seemed fettered in word or deed'.[12]

This select band of owners was offered two options. Under the first scheme they could hand over their houses to the Trust. Under the second scheme they could be affiliated in some unspecified way in the hope that government tax concessions would in due course be forthcoming. In both cases the owners would have to agree to open their doors to the public for thirty days a year. Included in the letter was a brochure outlining certain tax advantages in taking up the first scheme and possible advantages of the second, along with an offer of a visit from the country-houses secretary if they were in any doubt as to the schemes' advantages or in need of further elucidation or persuasion. It was in this role that Lees-Milne was central to the eventual success of the operation.

Lees-Milne was born in 1908. After leaving Eton in 1926 he was told by his eccentric father that the days of privilege were over and was taken to a London employment agency called Useful Women where he sat with twenty girls learning how to type. Eventually, with help from his mother, he went to Magdalen College to read history, which he found a waste of time. However, Oxford instilled in Lees-Milne a passion for architecture, which was reinforced by an evening when he had gone to dine at Rousham, a Jacobean house on the Cherwell between the city and Banbury. After dinner the host became:

> noisy and rowdy. On leaving the dining-room he got hold of a hunting crop and cracked it against the portraits. With the thong he flaked off chunks of paint. When satisfied with having worked off some of the effects of his brandy on the Knellers and Reynoldses, he fetched a rifle from the gun-room. He went to the terrace and proceeded to fire at the private parts of the statues . . . . I felt numb with dismay and

84

misery. The experience was a turning point in my life. It brought home to me how passionately I cared for architecture and the continuity of history, of which it was the mouthpiece.[13]

After working for Lord Lloyd, the imperial administrator, and then for the chairman of Reuters, Lees-Milne found himself in 1936 without a job. A chance meeting with Vita Sackville-West, who was a close friend of Matheson's sister, brought him an interview at the Trust and he was duly appointed as the first country-houses secretary. His initial stint lasted from 1936 to 1939, when he joined the army. He returned to the Trust having been invalided out of the army, and stayed as secretary until 1950 when he became historic-buildings adviser, a part-time post which he held until 1965. Finally he became a member of the Properties Committee, a descendant of the Country Houses Committee, until he retired in 1983. His diaries belittle his own contribution, but without his energy, his enthusiasm and his sympathetic understanding of the owners' difficulties, the Country-House Scheme might well have died at birth. His strong historical sense and his appreciation of architecture enabled him to give direction to the instincts of the leaders of the Trust.

A number of owners took up the offer of advice from the secretary of the Country Houses Committee, and Lees-Milne set off on the first of his many visits to a succession of historic houses. Travelling by train, by bicycle or on foot, this summer of 1936 was to be the most enjoyable of his life and a good deal less arduous than his experiences in the icy winters of the 1940s, which he recounts in his diaries. Still only twenty-eight, he was almost invariably received with friendliness and hospitality, though his sang-froid was occasionally put to the test. At the conclusion of a fruitless visit to Longleat, Lord Bath ordered his guest's motor car to be brought round and insisted on accompanying him to the door.

The steps to the drive were flanked on either side with a row of footmen in livery. In place of my uniformed chauffeur an extra footman wheeled my bicycle to the front of the steps. I shook my host's hand, descended the perron and mounted. At the end of a straight stretch of drive, having turned smartly to the right, I looked back for a last view of the glorious façade. Lord Bath, attended by his posse of open-mouthed and doubtless disdainful servitors, was in the old-world manner of true hospitality still standing at the top of the steps until his guest was out of sight. I gave a half-hearted wave, and pedalled ahead.[14]

As Lees-Milne admits, the Country-House Scheme did not start with a bang. Neither the owners nor the Treasury liked the second scheme and this was, therefore, abandoned. The response of the owners to the first scheme was, however, less hostile, and the Treasury suggested that the Trust should be encouraged to proceed with it.[15] The deal on offer was a simple one. A donor would give his house with an endowment in the form of land or investment and in exchange he and his heirs would continue in occupation. The house would be declared inalienable and thus could never be sold or mortgaged. Like Lord Lothian, the Trust attached great importance to the family continuing to live in the house. Originally there was no general rule relating to the terms whereby future generations would occupy the house, each case being subject to separate negotiations. The rule that only the donor and his heir could live rent free and that thereafter a market rent would be charged was not introduced until 1970, after criticism of the arrangements being too beneficial to some of the donor families: how beneficial depended on the size of the endowment.

The Trust had learnt from its experience at Montacute and elsewhere that a house cannot be supported out of income from visitors alone and that an endowment was required to keep the property in good order. Initially the endowment was calculated on the basis of the sum required to bring the building up to date, on the assumption that, should any house fall vacant, it could be let if suitably modernized. This seemed a reasonable solution at a time when it was usual for large country houses to be rented for several years or more. The other lesson learnt from Montacute, which the Trust had accepted with only a few family portraits and Lord Curzon's bath, is that a house without contents is no more than an empty husk. Although some owners were willing to give the furniture and pictures to the Trust, the more usual arrangement was for the contents to be left on loan. For houses such as Knole, which came on long lease under the 1939 National Trust Act, official permission could not legally have been obtained for the contents to be made over to the Trust.

In April 1937 the National Trust Act passed unopposed through both Houses of Parliament, providing the basis for the Country-House Scheme and giving the Trust power to hold land and securities for the upkeep of its property. The donor might continue living in the house without being taxed on the amenity, subject to allowing access for the public under a memorandum of wishes or a clause in the lease. The purposes of the Trust were enlarged to include responsibility for land surrounding a house and for 'the preservation of furniture and pictures and chattels of any description having national, historic or artistic interest', and the definition of buildings the Trust might hold was extended from those of 'historic interest' to include those of 'architectural or artistic interest'. The Act empowered local authorities, with the consent of the government, to vest land or buildings in the Trust, and to contribute to the acquisition and maintenance of Trust properties. Several local authorities later made use of this power and thus enabled the Trust to take on important houses which did not have adequate endowments. Finally the Trust was empowered to accept restrictive covenants without the necessity of having an interest in adjacent land. This was an encouragement to people who wanted to safeguard their estates from development without divesting themselves of ownership. In addition, the restriction on development reduced liability for death duties and there was no obligation to provide access.

No sooner had the 1937 Act been passed than the Trust realized that it had made no provision for settled land left to named successors. This means that the owner of an entailed estate was a tenant for life and could not leave or sell it to whom he pleased, even if it appeared to be in the interest of the family for the property to be passed to the Trust. In total it was estimated that there were no more than about 300 entailed estates in Britain, but they included Knole and other important houses. A second Bill was therefore drafted to enable the Trust to take on such estates. This Bill, however, ran into difficulty when the Lord Chancellor, Lord Maugham, considered the matter important enough for it to be presented to a select committee rather than following the usual procedure for Private Bills. Although, as it turned out, the Bill, which became law as the National Trust Act 1939, has been brought into use on only a handful of occasions, it was strongly opposed by several peers on the ground that it would allow an entailed owner 'gratuitously or otherwise' to make his estate over to the National Trust without the consent of his successor or trustees.

Lords Lothian, Esher, Crawford and Methuen spoke on behalf of the Trust, but it was Lord Zetland, in his role as chairman, who delivered the most effective speech.

> It is not the National Trust which is disinheriting remainder men [successors], it is the state itself. It is quite impossible, under our present system of taxation, for the owners of properties of this kind to maintain them. The whole object of the National Trust is to enable [owners] to see their residences and their grounds and properties maintained and to occupy them for life. This is no small effort on the part of a body of men, the National Trust, who for forty-five years

have been working, giving their services, their thought, free, gratis and for nothing, to save some of the historic buildings of this country – the pride of our English life – for the people of this country.[16]

Much of the impetus behind the Country-House Scheme came from Oliver Brett, third Viscount Esher. He had joined the Executive Committee in 1926, and became chairman of the Finance and General Purposes Committee in 1935. From an early age he had developed an interest in architecture, encouraged by his courtier father, who had been a confidant of both King Edward VII and King George V. He was an active Liberal, but he devoted most of his public life to advancing the arts and defending fine buildings and countryside, not only through his work with the National Trust but also as chairman of the SPAB, and of several organizations connected with the theatre. The sharply discriminating Lees-Milne, who worked under him for many years, was an almost unqualified admirer. 'To be in his company was a delight, to witness his handling of committees was an education. With infinite merriment he always got what he wanted and what he wanted was invariably right. . . . He was a shrewd, genial, irreverent and witty man – in fact the funniest I have met.'[17] Small and quizzical, he resembled Hercule Poirot in appearance. Christopher Gibbs, who was later chief agent, remembered Esher for his wit, but also for his ability to wound.

Lees-Milne's task was not an easy one – to persuade the owners of some of England's greatest houses to present them to the National Trust, still a relatively small organization, along with an endowment to keep them going. His task needed diplomatic skill, the ability to command the confidence of potential donors and even an occasional sleight of hand. The owners themselves were often in grave financial difficulty and torn over what to do with the crumbling piles which they could no longer afford to run in the way to which they had been accustomed. All the Trust could offer, if an owner did decide to hand over his property, was continuity of tenancy, relief from death duties and a promise that the Trust would look after it.

The Country-House Scheme would not have succeeded had Lothian proved wrong in thinking that a considerable number of owners would be sufficiently concerned to see their houses preserved that they would give them to the Trust. This was not an attitude that would have been easily understood on the continent. But because Britain had been spared the disruption of war and revolution, many large houses had remained within a single family for centuries, and because of the practice of primogeniture many were still supported by agricultural estates. Their owners often felt themselves to be in the position of trustees with a duty to pass on their estates to future generations, a belief fostered by the rules governing entailed estates. These influences are well described by Nigel Nicolson, a cousin of the fourth Lord Sackville, the donor of Knole in Kent, who was himself responsible for the transfer of Sissinghurst with its garden created by his parents: 'Childhood associations with a single place, the slow revolution of generations of the same family and their servants, the very portraits on the wall which gave the house a second and watchful lot of occupants – all this created a gentle momentum which safeguarded the house and made sacrifices on its behalf entirely natural.'[18]

Thus when death duties or other financial circumstances made it difficult to carry on, some owners – rather than see the house converted to institutional use, its contents dispersed and the park sold for building – were glad to hand over to the Trust. Sometimes there were also personal factors, particularly the lack of a son with whom to share the burden. This was the case in about 40 per cent of the houses transferred to the Trust and was undoubtedly the deciding factor for Sir Henry Hoare, the owner of Stourhead in Wiltshire, the first house to be offered under the scheme, even before the legislation had been enacted (though it did not in fact come to the Trust until after

Hoare's death in 1947). The Hoares had lost their only son during the First World War. They were elderly and very pre-occupied with the future of the house and its superlative garden, to which they had devoted their lives. Lees-Milne gives an amusing but touching picture of the couple and their difficulties in coping with the rigours of the 1939 and earlier wars:

Sir Henry is an astonishing nineteenth-century John Bull, hobbling on two sticks. He was wearing a pepper and salt suit and a frayed grey billycock over his purple face. He had a very bronchial cough and kept hoiking and spitting into an enormous carrot-coloured handkerchief. . . . Lady Hoare is an absolute treasure, and unique. She is tall, ugly and eighty-two; dressed in a long black skirt, belled from a wasp waist and trailing over her ankles. . . . She has a protruding square coif of frizzy grey hair in the style of the late nineties, black eyebrows and the thickest spectacle lenses I have ever seen. She is nearly blind in one eye. She is humorous and enchanting.

We ate in the little dining room at a long table, Sir Henry with his back to a colonnaded screen, Lady Hoare with hers to the window. He spoke very little, and that little he addressed to himself. She kept up a lively, not entirely coherent prattle. She said to me, 'Don't you find the food better in this war than in the last?' I replied that I was rather young during the last war. 'Oh,' she said, 'you were lucky. We were reduced to eating rats.' I was a little surprised until Sir Henry looked up and said, 'No, no, Alda. You keep getting your wars wrong. That was when you were in Paris during the Commune.'[19]

They both died on the same day in 1947.

The first house that was actually handed over to the Trust under the Country-House Scheme was ironically not some grand mansion set in a landscaped park and belonging to a noble family, but a Victorian house just outside Wolverhampton, which had been built by an enlightened industrialist only a few years before the Trust was founded. Wightwick Manor was designed by a disciple of William Morris and its interior reflects the craftsmanship of his associates. The property belonged to Sir Geoffrey Mander, a Liberal Member of Parliament, who had inherited the paint business and the house from his father. On 10 August 1936 he wrote to Matheson explaining that he had been staying with Sir Charles Trevelyan, the brother of the historian, who had told him of his plans to give Wallington, his Northumbrian estate, to the Trust and this had caused him to wonder whether it might be possible to do something of the same kind. Matheson replied discouragingly that, while Wallington was a place of historic interest within the meaning of the 1907 Act and much of the estate qualified as of natural beauty, he doubted whether the same view would be taken of Wightwick. Mander persisted throughout a long correspondence and finally Lady Trevelyan, wife of Sir Charles and by now a member of the Executive Committee, persuaded her colleagues to accept. Rosalie Mander, Sir Geoffrey's second wife and biographer of Rossetti, continued to add to the collection of works by the Pre-Raphaelites and by William Morris and his associates with which to embellish the house.

Sir Charles Trevelyan had been an advanced Liberal MP up to 1918. He then joined the Labour Party and was a Newcastle MP from 1922 to 1931 and President of the Board of Education in the two minority Labour governments. A month before Lothian's historic speech he had notified Matheson of his intention to give Wallington with its 13,000 acres to the Trust. Three years later, after lengthy discussions with officers of the Trust, including his brother, who was chairman of the Estates Committee, he explained his reasons in a radio broadcast:

> In every part of England now there are well-known country
> houses which are ceasing to be occupied by their owners.
> Their contents are being sold and with them departs the
> historic continuity of the house and its principal interest and

value. . . . Now I do not mean to say that in every case where this is occurring throughout England there is a great public loss. In some cases it is only a misfortune for the owners . . . but in others the community loses as much in the long run as the private owner, where treasures of national interest are dissipated and the fine architectural or historic mansions are doomed to decay.

Charles Trevelyan, who had for some time opened the house and the grounds to the public, continued:

It became evident to my wife and myself that it would be a great public loss if Wallington could not continue to be open for the recreation of the increasing numbers of people who now visit it. It would clearly be a public calamity if the house were to lose the collections which make it interesting, or were to be closed to the public who increasingly like to frequent it, or if the woods were cut down to meet taxation, of if the place were to become derelict if my successors could not afford to live in it. I made up my mind that I was not going to allow Wallington to become a memory.

His only condition in giving the house to the Trust was that his wife and children could live there during their lives; thereafter 'the Trust will not be bound to let Wallington to Trevelyans, but will be free to let it to whoever they think will be the best guardians in the public interest'.[20]

In 1942, when the estate was transferred to the Trust, he put a further gloss on his motives in a notice issued to the press: 'Sir Charles Trevelyan is a socialist and believes it would be better if the community owned such houses and great estates. He was also influenced by Lord Lothian with whom he discussed the whole question some years ago.'[21] Wallington was a singularly appropriate place to come to the Trust since it had for many decades been a meeting place for public servants, scholars and writers, including Ruskin, who had painted flower decorations for the central hall. The gift in one bound increased the area owned by the Trust by an acreage equivalent to the total holding in 1924.

Miss Matilda Talbot, of Lacock Abbey in Wiltshire, was another owner who saw herself as a guardian for the nation. In September 1938 she wrote to Matheson expressing interest in the Trust's new scheme. Unmarried, she had inherited Lacock from an uncle in 1916 and wanted to transfer not only the house, which she had already opened to the public, but also the adjoining village and about 500 acres of farmland. In his initial report on the property, Lees-Milne notes, 'Miss Talbot wanted her niece to live there after her death if she could afford it, but her chief motives appear to be political ones for she is of the opinion that the public should have as much right to enjoy Lacock as herself.'[22] If the family did not wish to live there she wanted the house to become a religious institution for women. This would have been appropriate, for the Tudor house had been built out of a medieval nunnery. It had the additional interest of being the place where the first dim print was produced by the pioneer photographer William Henry Fox Talbot. The valuable library had been sold by Miss Talbot in order to provide the village with electricity.

Negotiations for the transfer lasted nearly seven years. The difficulty with Lacock, as was to be the case with many other properties, was the endowment. If Miss Talbot were to give her estate of 500 acres as well as the village she would have nothing left to live on; she was also concerned about the position of her heir. Several owners were already toying with the scheme and Matheson felt that one successful example would lead others to follow suit. He wrote a number of letters to Miss Talbot, pressing upon her the importance of her gift and emphasizing that speed was of the essence. But in July 1939 Miss Talbot broke off negotiations. Perhaps she felt too harassed, perhaps the Trust had been too eager. Whatever the reason, she wrote to Matheson with some advice:

You must always face the likelihood of changes in the political complexion of our government and I have been disturbed to feel that the Executive Committee of the Trust, which used to include people of all shades of political opinion, have been getting rather more staunchly Conservative. This might react adversely on the Trust, particularly because of the danger of the Country-House Scheme being thought by some people to be a scheme for the advantage of the landlord – particularly so if there were a strong Labour government.[23]

These were perceptive comments. Although the social composition of the Council, half its members being elected and half nominated, had changed little, there was already some change among the leaders of the Trust and this became still more marked after 1945. It was a change which went against the socio-economic national trend. Put bluntly, the Trust became more upper class. Those who had known the founders of the Trust were inevitably giving up: Harriot Yorke, who represented the Octavia Hill point of view on the Executive Committee, though she seldom intervened in discussions, ceased to be treasurer in 1924 and died in 1930; Nigel Bond, who had been secretary from 1901 to 1911, retired as chairman of the Estates Committee in 1928; and Stenton Covington, who was a powerful member of the Executive Committee and started the Trust on its acquisition of the coast with his gifts near Fowey, died in 1935. Admittedly there was still a link through Janet Upcott,* who had worked with Octavia Hill and remained on the Estates Committee until the late 1960s. Until 1964 there was also a link through Sir Robert Hunter's daughter, Dorothy, who had followed her father on the Executive Committee and was always listened to in respectful silence. Together they served consecutively for almost seventy-one years.

*In 1927 she was appointed the first woman housing manager to a local authority by Parker Morris (father of Jennifer Jenkins), then town clerk of Chesterfield.

The 1930s, however, had seen Lord Zetland take over from John Bailey as chairman, and Lord Esher from R. C. Norman as chairman of the Finance and General Purposes Committee. G. M. Trevelyan continued as chairman of the Estates Committee until 1949, when he was succeeded by the ninth Earl De La Warr. Looking back when he wrote his memoir, Chorley thought that 'the aristocratic tinge' which had always existed in the Trust had been accentuated after the Country-House Scheme was introduced. Despite his own earlier hostility to aristocrats, which he says largely evaporated as a result of his experience at the Trust, he believed that 'the knowledge, the loving familiarity of so many members . . . who would have been in the habit of spending long periods in visits to our great country houses is of inestimable value to their protection and sensitive management and could be obtained in no other way'.[24]

In 1943 Miss Talbot reopened negotiations. This time Lees-Milne was in full charge and he visited her in December, noting in his diary: 'She is a dear, selfless woman, and extremely high-minded. She has the most unbending sense of duty towards her tenants and the estate to the extent that she allows herself only a few hundreds a year on which to live. She spends hardly a farthing on herself and lives like an anchorite.'[25] On his next visit, Miss Talbot gave him goose for lunch, a rare delicacy during the war. Whether or not he was influenced by this he records: 'I think Miss Talbot is one of the noblest and most exemplary benefactors we have had to deal with.'[26] By January 1944 agreement had been reached. The Trust would take the Abbey, the village and 300 acres as endowment and Miss Talbot would have the rest of her estate to live on.

When war became imminent the Trust moved its head-quarters out of London to two tiny pavilions at Runnymede. But these were obviously too small, and the day after war was declared the office opened in the classical but chilly splendour of West Wycombe Park, which was at the time subject to negotia-tion and which led to it being given to the Trust by Sir John

Dashwood in 1943. Executive meetings were held at irregular intervals and in December 1940, after an 'unpleasant experience', Zetland insisted that they should not coincide with a full moon. Esher was often the only man available for committees and held the meetings at his own house, Watlington Park, not far from West Wycombe. Essential business was conducted by Esher and Matheson, but otherwise the first eighteen months of war were a dead period. The situation was summed up by Norman in a letter to Esher: 'We are very fortunate in having you so handy to the National Trust's war-time offices. I repose the fullest confidence in your totalitarian regime.'[27]

At the end of 1941 James Lees-Milne again became available as Country-Houses secretary. Matheson, who had been corresponding with several potential country-house donors, wrote to him: 'We are now fairly flooded out with work on the subject as a considerable number of owners have asked us to put schemes before them, some of them owners of entailed estates. I think that the trend of taxation resulting from the war has undoubtedly helped many to see more clearly the advantages of coming into the scheme.'[28] His predictions were not far wrong, but there were problems. Lord Esher wrote to the Executive Committee in February 1942: 'I am somewhat disturbed by the failure of the Trust to obtain any considerable number of country houses, and by the apparent difficulty of reaching agreement between the Trust and those owners who have shown an inclination to take advantage of our services.'[29] His view was confirmed a few months later when Earl Spencer broke off negotiations about the future of Althorp, writing: 'I do not think many people will hand over their ancestral property until the Act of Parliament is made more attractive to owners.'[30]

While Lees-Milne had been away three houses had passed to the Trust in addition to Wallington: Packwood, a much restored Tudor manor house with a famous garden of yew trees near Birmingham; Cliveden, with its magnificent park and hanging woods plunging down to the Thames, where Lord and Lady Astor entertained Lord Lothian and other public figures in the 1920s and 1930s; and Blickling, which came to the Trust after Lothian's death, one of the most beautiful country houses, containing a fine library and accompanied by an estate of 4700 acres, then in a rather dilapidated condition.

Although Cliveden was presented with the largest endowment that had ever been provided for a Trust property, for many others the question of an endowment was a real difficulty, as it has continued to be. If the funds could be raised, the owner might be left with nothing to live on. Esher warned the committee that unless they were prepared to accept properties with a minimum endowment they would not receive many offers. He suggested that it should be the families' responsibility to maintain their houses at a standard to suit their inclination and purse and, if they left, the Trust need merely keep the roof watertight and the dry rot at bay until a tenant was forthcoming. His proposal was agreed and a number of houses came to the Trust with insufficient endowment. In 1943 there were two: Great Chalfield Manor in Wiltshire with only 9 acres and West Wycombe Park with 338 acres, including 38 which were to be left alienable in the hope that development value would provide for an endowment.

One of Lees-Milne's first tasks on returning from the army was to deputize for Matheson, who had been taken ill. In February 1942 he notes in his diary, 'Rather nervous how I was going to acquit myself this first time I have ever taken a meeting alone. But as a matter of fact all went fairly well. The old gentlemen are so fearfully ignorant of the intricacies of most items on the agenda that one need have little fear of them. Besides, they one and all are so nice.'[31] The early diffidence was clearly evaporating. As G. M. Trevelyan pointed out, if Lees-Milne had not been invalided out of the army when he was, the Country-House Scheme, with Matheson being ill much of the time, might never have got off the ground. With more owners asking for advice, Lees-Milne set off round the country with the National Trust car and Eardley Knollys, his companion on many of these visits. Knollys had been appointed assistant agent in 1943 on the strength of his wartime experience as an agricultural worker. Before the war he had owned a gallery in Knightsbridge, he was a friend of Graham Sutherland and Duncan Grant and his real interest lay in early twentieth-century painting. Not surprisingly he did little to learn about estate management, as Matheson was soon complaining.

In 1943 the Trust moved back to London, and from March 1944 Lees-Milne acted as secretary while Matheson was away on six months' leave. Lees-Milne notes in his diary that they were days of 'feverish activity in the office'. On one occasion he had to go through the Executive Committee agenda with Zetland. 'He was remarkably friendly and to my amazement some item amused him and he began giggling. Then he pulled himself together with a jerk. But it was too late. He had betrayed the fact that he was a human being. I shall no longer be in awe of him.'[32] At another meeting Lees-Milne records, 'We had a Country-Houses Committee in the afternoon. Every time I thought I heard a buzz bomb I warned Esher, who is slightly deaf, and without shame he threw his papers before him and dashed to the door and the staircase, then laughed at himself.'[33]

The most pressing issue for the Trust by 1944, a year in which seven country houses were taken on, was what would happen if the families no longer wanted to live in them. Early in 1943 Zetland had written to *The Times* to find out whether there were any public institutions anxious to establish themselves in large country houses. In the ensuing year, although there had been several enquiries, none had come to fruition. The reasons were partly the difficulty and cost of adapting buildings to uses for which they were not designed, and partly the difficulty of reaching a compromise over which parts of the house could be occupied. What the Trust wanted to avoid was the gradual metamorphosis of the country house into the country museum. The *Daily Mirror* summed up the Trust's feeling in a leader which stressed, 'At all costs we must avoid a stuffed England.'[34] The Trust was acutely aware of the problem and the annual report of 1944 notes, 'English country houses were built to live in, and the purpose for which the National Trust exists is to preserve their character unimpaired; no consideration must be allowed to detract from this policy.'[35]

One house which was no longer lived in by its owners was Speke Hall on the banks of the Mersey, only a few miles outside Liverpool. Built over a century from 1490 and altered internally by the Victorians, it is one of the best examples of black-and-white half-timbered mansions in England. In 1796 it had passed to the Watt family and the last person to live in it, Miss Adelaide Watt, had died in 1921. In her will a clause had been inserted which empowered the trustees to convey the house with its contents and grounds to the Trust. This had been overlooked and in May 1937 the SPAB wrote to Matheson warning that Speke, listed by the Trust as being among the 230 special houses, was threatened with demolition. But it was not until August 1942 that a letter was received offering Speke to the Trust. In the absence of anything like adequate endowment, the Trust decided to ask the Liverpool City Council whether they would be interested in taking a ninety-nine-year lease on the house. Discussions lasted until 1944, when the City decided to accept the Trust's offer. Later the Merseyside Metropolitan County Council took over the lease until in 1986 the house was returned to the Trust with a sum of £90,000 towards repairs and an annual payment of £100,000 for thirty years in recognition of the unexpired portion of the lease; although not a full endowment, this was a better deal for the Trust than the initial gift.

Other houses taken on during 1944 included three of disparate history and character. Gunby Hall in Lincolnshire is a classic Queen Anne mansion set in rich and well-tended agricultural land. The house had been in the same family since it was built, but Lady Montgomery-Massingberd and her husband had no son, and were worried that the Air Ministry's plans for the adjacent aerodrome would endanger the trees and even the house. They thought that the most effective way to ward off the immediate and future threats would be to hand over to the Trust without delay; this was done and Lees-Milne records that the transfer of a large estate had never proceeded more speedily or smoothly.

The other two houses had been bought by their donors. The first was Lindisfarne Castle, romantically situated on Holy Island in Northumberland. It had been transformed by Sir Edwin Lutyens for Edward Hudson, the founder of *Country Life* and other periodicals. Some years later it was bought by the merchant banker Sir Edward de Stein, who gave it to the Trust.

The second was Polesden Lacey in Surrey, which was bequeathed with an endowment and 900 acres of park and woodland by Mrs Ronald Greville. She had bought the house in 1906 to display her outstanding collection of works of art and as a venue to entertain the prominent literary and political figures and members of ex-royal families who adorned her weekend parties.

Towards the end of 1944 the Trust was still taking a pessimistic view of the way the Country-House Scheme was progressing. Since 1937 the Trust had acquired twenty-three houses, not all under the Scheme, of which only a few were of first-class architectural importance, though some had outstanding works of art or parks. A further twenty-six houses had either been offered or had already fallen through. Only three of these eventually came to the Trust: most of the rest have since been sold up and have lost their historic contents. An unsigned paper in the Trust archives entitled 'A Review and a Forecast', perhaps written by Matheson, notes that there was some public apprehension that the Trust might allow donors' families to have an influence on their management and expresses a doubt whether the scheme was on a long view financially sound. The memorandum concludes that the Trust should attempt to find a solution on the lines of the agreement with Liverpool for other houses.[36] Lees-Milne, writing at much the same time, was convinced that the Trust was reaching deadlock in the case of country houses offered by owners who had no intention of living in them again. 'These owners in offering their houses (and possibly certain valuable contents, gardens and grounds) unconditionally to the Trust not unnaturally object to being asked to provide an endowment sum for upkeep, from which in the future they and their descendants will derive no financial benefit at all.'[37]

One such was the Duke of Wellington, who announced that he would give Apsley House in London to the government rather than to the Trust since the large sum required for an endowment could then go to keeping up his country house, Stratfield Saye, and he would still be able to keep rooms at the top of Apsley House.

Many owners were becoming increasingly doubtful whether they would be able to live in their houses after the war, and few envisaged residence beyond one or two generations. Lees-Milne also thought that the answer might lie with public institutions and meanwhile suggested that the Trust might ask for government funds to preserve houses from which only the public would benefit.

The one house of significance to come to the Trust during 1945 was Hatchlands in Surrey, which belonged to Hal Goodhart-Rendel, an architectural historian, who had inherited it from his grandfather. Built in the 1750s the house has the earliest-recorded interiors by Robert Adam and is surrounded by a park designed by Repton. When the war against Japan ceased on 14 August 1945, negotiations for several important properties were nearing completion. One of these was Petworth in Sussex, which John Wyndham, the nephew and heir of Lord Leconfield, was anxious should be given to the Trust. The arguments used in persuading his uncle were those that Lees-Milne had already found convincing:

> We decided to stress upon the feudal, reactionary Lord L that 1) during his lifetime he will not be disturbed in any particular, 2) he may be better off financially, 3) he may retain the contents (for the best will be exempted on J[ohn] W[yndham]'s succession) and 4) by transferring now he will establish the only assurance that his successors can live at Petworth.[38]

Lees-Milne's flippant entry in his diary for 1 June 1945 cannot disguise the crucial role he played in persuading owners that it was in their own interest to transfer their houses to the Trust:

> The lengths to which I have gone, the depths which I have plumbed, the concessions which I have (once most reluctantly) granted to acquire properties for the National Trust, will not be known by that august and ungrateful body. It might be shocked by the extreme zeal of its servant if it did. Yet I like to think that the interest of the property, or building, rather than the Trust has been my objective. I have to guard against the collector's acquisitiveness. It isn't always to the advantage of a property to be swallowed by our capacious, if benevolent maw. These pious reflections came to me in the bath this morning.[39]

CHAPTER FIVE

Subtopia and the Second World War

D ESPITE THE importance of the Country-House Scheme for the Trust, in the 1930s public opinion was more con-cerned with threats to the countryside than to historic houses. 'Nothing can stop the destruction of our countryside by town influences,'[1] lamented A. G. Street, the Wiltshire farmer and writer, as he watched the impact of private motoring and regular bus services on land which until then had remained relatively undisturbed. Between 1927 and 1939 hundreds of thousands of acres were built over for leafy suburbs, spreading out from the densely packed towns, for retirement homes blanketing the south-east coast, and for the expanding con-sumer industries and services. The clear distinction between town and country was becoming blurred.

The Town and Country Planning Act, which had been introduced in 1931 with the hope of controlling this develop-ment, proved ineffective after it had been emasculated when the National Government came in later that year. Nor did the campaign to protect some of the most beautiful landscapes by establishing National Parks achieve its goal. But two more limited measures demonstrated that something could be done to resist 'town influences'. In 1935 the Restriction of Ribbon Development Act checked the continuous lines of villas being built along the new arterial roads and three years later the Green Belt (London and Home Counties) Act led to the protec-tion of some 35,000 acres round the metropolis.

Town dwellers had more leisure to spend in the country-side, particularly after 1938 when a week's statutory holiday with pay was introduced. Transported by car, charabanc or bicycle they could penetrate even the more remote areas. Their numbers lent strength to the ramblers' lobby, which in 1939 at last secured an Access to Mountains Act. This was the last of a series of Bills introduced since 1888, but it did not achieve the general right of access to uncultivated land which was the ramblers' objective. The newcomers to the countryside were not without somewhat heavy-handed instructions about how they should behave when they got there. A pamphlet entitled *The Complete Hiker and Rambler* advised, 'Do not put on superior airs when talking with farm or other country folk. They are far from being "simple". Don't be annoyed if they show natural curiosity as to your camping arrangements. There is often little else for them to be interested in.'[2]

The mounting pressure on the countryside was undoubtedly one factor in the increased flow of gifts in land and money to the Trust. In the years 1935–40 some 19,000 acres were acquired, the highest total for any five-year period up to then. But it was in 1940–5 that the rate of acquisitions really took off, reaching over 43,000 acres in that quinquennium and not falling much below that figure apart from the years 1970–5. In this long-term expansion the large parks and estates transferred to the Trust under the Country-House Scheme formed one important element.

While protecting the Lake District had always been a primary objective of the Trust, the Peak District had been ignored. The Trust did not own a single property in the area until 1930 (when an anonymous gift of a nine-acre hilltop above Wirksworth was received) and by 1945 it owned only 2500 acres there compared with 15,000 in the Lake District. However, by the 1930s the Peak District was in even greater need of protection than the Lakes. One look at a map shows that Dovedale and the Manifold Valley, the location of two of the Trust's earliest acquisitions in the area, are dangerously close to the conurbations of Yorkshire and Lancashire. Today the Peak National Park is within a fifty-mile radius of more than one-third of England's population, with Manchester, Sheffield and Leeds all within easy reach.

In the early 1930s it was estimated that nearly 15,000 people left Sheffield and a further 15,000 left Manchester on an average Sunday to escape from the smoky atmosphere of the slums into the pure air of the Peak District. But unlike the Lake District, where most of the fells were by long-established custom open to all, in the Peak District practically all the moorlands were closed to the public. Nor were there many footpaths. In an area of 215 square miles in and adjacent to the Peak District, there were only twelve footpaths that exceeded two miles in length.[3]

Benny Rothman, then the twenty-year-old secretary of the British Workers' Sports Federation, recalls his members' increasing frustration at coming out into the country and then being unable to enjoy the freedom of the moors or to get to the top of the hills.[4] At Easter 1932 he and half a dozen others, including some visitors from London, were turned back by a group of gamekeepers. They returned very upset, but they reckoned that if a lot more of them went out they would outnumber the keepers, so they decided to organize a mass trespass. Their plan attracted a lot of publicity. When the day came, Benny Rothman was astonished to see that hundreds of ramblers had turned up and that there were also a large number of police. The ramblers moved off towards Kinder Scout, the highest point in the area, on which most of the local ramblers had never set foot. Rothman recounts what happened: 'We started to slowly scramble up a very steep bank and we got two-thirds of the way up when we suddenly heard a commotion and a group of gamekeepers came charging downhill waving their sticks and shouting "Get back". There must have been about a dozen keepers and 400, 500 or even

600 of us so it didn't make much impression on us. Two scuffles took place.' The ramblers went on to reach the top of Kinder Scout. When they returned six were arrested and later charged with riotous assembly and other offences. Benny Rothman and four others were sentenced to periods of up to six months' imprisonment. The mass trespass became a *cause célèbre* for the ramblers' movement and a significant event in the campaign for access to the uplands.

While the Trust realized that action was needed both to protect the landscape and to provide access for walkers, what it lacked was a Beatrix Potter, a G. M. Trevelyan or a Hardwicke Rawnsley to lead a campaign. That is until 1934 when F. A. Holmes, a businessman and philanthropist from Buxton, managed to persuade a rich Manchester merchant, Sir Robert MacDougall, to buy Ilam Hall in Dovedale and present it to the Trust for use as a youth hostel.

From the grounds of Ilam Hall a visitor can see the bare hills that form the first miles of the Pennine Way running along the spine of northern England. Two centuries before, Dr Johnson had been there and could not believe that the Manifold River he saw near the house was the same as that which sank beneath the ground five miles away, even though, as Boswell recounts, the gardener of the Hall 'said he had put corks where the Manifold sinks into the ground and had caught them in a net placed before one of the openings where the water bursts out'.

F. A. Holmes had a great knowledge of the Peak District and of Dovedale in particular. Once when he was a boy walking through the woods of the dale, so the story goes, he heard the trees being felled for use in the First World War and from then on he was determined to protect the unspoilt valley. After the war he frequently wrote to S. H. Hamer demanding to know what the Trust was doing to safeguard it. But until he took action himself the answer was very little. After that, with the help of Sir Robert MacDougall and the Pilgrim Trust, the

Trust's acquisitions in Dovedale and the Manifold Valley gathered pace. They included famous rock formations with names such as Jacob's Ladder and the Twelve Apostles as well as several farms and outlying hills. Other early acquisitions in the Peak District included a group of farms in Edale typical of the Pennine country – rolling moorland capped by clumps of trees and steep hillsides falling to winding river valleys; Mam Tor or the shivering mountain, so called because of continual landslips over the years; and part of the Longshaw estate only eight miles south-west of Sheffield. Fifty years later the massif of Kinder Scout was bought by the Trust, with the help of a public appeal and government grants, to form part of the Peak Distrct estate, now totalling 36,000 acres. Co-operation between the National Park and the landowners, of which the National Trust is the largest, allows grouse shooting to coexist with access to the uplands.

For coastal conservation a growing impetus came not from one or two influential campaigners but from a host of individuals and groups. During the decade before the war more than thirty small seashore properties were given to the Trust in Devon and Cornwall alone, either freehold or under covenant. On the Lizard Peninsula, for example, three owners gave part of Kynance Cove in order to stop the spread of squalid shacks and scattered houses which by 1935 had disfigured the Point two miles to the south. Public anxiety was echoed in the press. Two years later the *Daily Herald* wrote:

> Hardly a government in Europe would permit private individuals to disfigure and destroy the natural beauties of country or coast as they have almost unrestricted freedom to do in Britain, the loveliest of all. The Danes have just passed a law restoring access to the whole coastline and forbidding all building within 100 yards of the shore except by direct permission of the State. The British coastline wins the praise of the world. How long shall we wait and how much will be left to save when we copy this civilized Danish land?[5]

It was the Trust not the government which answered this call. The Trust had recently acquired two dramatic headlands in Pembrokeshire, a county which had hitherto been protected by its remoteness from the bungalows and caravans which were spoiling much of the north Wales coastal strip. With the encouragement of Tom Jones, secretary to the Pilgrim Trust and a confidant of Baldwin, the Trust decided to try and preserve as much as possible of the coast between St David's Head and Newgale, some fifteen miles to the south. There were in total forty different properties and the Trust engaged a land agent and a solicitor to negotiate with the owners. To pay for the project an appeal for £15,000 was launched in April 1939. The timing could not have been worse: by the end of the year, with just over £9000 raised, almost half from the Pilgrim Trust, plans were put on hold until after the war. Enough money had, however, been given to buy some, if not all, of the properties.

Faced with a yawning gap between its vision and its financial resources, the Trust tried the alternative route of protecting larger areas by means of restrictive covenants. A lead had been given by G. M. Trevelyan, who had bought land in the Lake District and on the Roman Wall, which he had then covenanted to the Trust. Starting with only a few thousand acres in 1930, by the outbreak of war 27,000 acres were protected by covenant, a counter to the criticism that the Trust was mainly concerned with preserving small beauty spots rather than with protecting wider areas of unspoiled beauty.

Covenants have proved valuable in preventing development but they cannot secure even essential repairs and, in areas of intense development pressure, it may require a great deal of time-consuming work to control small but cumulatively damaging changes. A good example is the 4000-acre Greenlands estate north-east of Henley-on-Thames, covenants over which were given in 1944 by Viscount Hambleden. Eardley Knollys reported at the time that the temple on Temple Island was in a deplorable state and that a programme of tree planting was needed. Since then those defects have been righted and the gift has ensured the survival of a particularly beautiful stretch of the Thames with its surrounding beechwoods, but it has entailed the detailed scrutiny of numerous applications to extend the many cottages and farm buildings.

An ambitious new move was made in 1934 with the acquisition of almost an entire village, West Wycombe in Buckinghamshire. More than twenty years earlier St Loe Strachey, the editor of the *Spectator* and a founder member of the Council, had proposed that the Trust should buy Pollington in Northamptonshire as a typical old English village, but on Octavia Hill's advice the idea had been turned down. 'Imagine the Committee faced with the prospect of turning out a poor widow,'[6] she commented. In the meantime experience had been gained in managing dozens of tenanted cottages round the country, and interest had been stirred by Sir Frank Baines, the government's Director of Works. In a lecture delivered to the Royal Society of Arts on 6 May 1926, Baines made an eloquent plea for the preservation of old cottages in villages and hamlets: these are records 'beautiful in themselves, of a culture entirely our own ... a documentation which once destroyed can never be replaced'.[7] Despite the difficulty of travelling on the fourth day of the General Strike, the audience assembled and with its support the Society proceeded to organize a conference, presided over by Stanley Baldwin, to raise funds for preserving villages.

Following this a few cottages were acquired and given to local archaeological societies, but nothing of outstanding interest was forthcoming until March 1929 when Sir John Dashwood decided to auction the village of West Wycombe in sixty separate lots. A week before the deadline, the Royal Society of Arts bought the shop, the post office, fifty-two cottages, three inns, two chair factories and a disused school. Four years later the RSA decided to offer the village to the

Trust for £25,000, having carried out extensive repairs and some internal modernization. The gradual improvement in the Trust's financial position made it possible to accept the offer, the transaction being made more attractive by the advice that it could be regarded as a sound investment in real estate. The village, which lies on the old road from London to Oxford, consists of a long main street flanked by a patchwork of flint, brick and timber buildings, many dating from the early sixteenth century. It was the first of many acquisitions of whole villages and hamlets. By 1990 there were almost sixty in the Trust's possession, yet this is an aspect of its work which often goes unnoticed.

The second village to come to the Trust was of a totally different character. This was Styal in Cheshire, which was created to house the workers brought in by Samuel Greg in 1784 to man the cotton mill beside the swift-flowing stream. The Gregs were a Unitarian family who over several generations provided not only neat rows of red-brick cottages but also an apprentice house, a school, two chapels and other amenities, all set against a well-wooded rural background. Finally in 1938, by which time the mill employed only a handful of workers, Alec Greg offered the village to the Trust: 'As one watches the growth of dreary unsightly villas all round greater Manchester I am more than ever impressed with the value of our property as a relief and contrast. And I should like the Trust to enjoy it as soon as possible.'[8]

Lacock, the third large village to be acquired by the Trust, came as part of Miss Talbot's gift under the Country House Scheme. It had been laid out on a grid plan under the aegis of the Abbey and its terraces of cottages, inns and bridges had changed little since the fifteenth century. Esher had earlier commented, 'I am afraid the beauty of the village may force us to take the house.'

By the mid-1930s the Trust was finding it difficult to maintain a close contact with its outlying properties. The management problems were particularly difficult in the Lake District, where the Trust's estate was growing rapidly. In the year 1933–4 alone, the Trust became the owner of an additional 1600 acres of land and water, including Buttermere, Crummock Water and Loweswater, as well as acquiring covenants over nearly 8000 acres from Wastwater in the west to Ullswater in the east.

To remedy this situation Bruce Thompson, who had been assistant secretary for eight years, was appointed the Trust's first representative in the north of England. This scholarly, self-effacing man, who had read archaeology at Cambridge, was the first of a succession of influential regional agents and directors who have made immeasurable contributions to enlarging the Trust's estate and raising the standards of management. 'Not a landowner, farmer or shepherd did not know and love him, and for thirty years or more he devoted his life to promoting the Trust's interests among the hills and fells of his native region.' The annual report for 1935 notes that 'He will be required to pay special attention to matters relating to the Lake District and there is every reason to believe that his presence at the Lakes will lead to the growth of the work of the Trust in this most important area.'[9] It was the first sign that the Trust's workload was becoming too great to manage from London.

In 1936 the first regional committee came into being. This was in Northern Ireland and was set up almost by accident. The Trust had been offered two small properties in the province, Killynether House and Ballmoyer House. Matheson had written to George Duggan, assistant secretary to the Stormont Ministry of Finance, to ask his advice on what action the Trust should take. The solution proposed was to set up a committee based in Northern Ireland to advise the Trust and look after any future properties in the region. Duggan and Matheson met to discuss its membership, and the executive in London duly accepted their nominations. Marcus McCausland, Lord Lieutenant for County Londonderry, was appointed chairman. He was a cousin of Christopher Gibbs, the Trust's land agent, who was sent to attend the first meeting and recalled that his cousin was 'violently anti-Catholic but agreed to have one Roman Catholic on the committee'.[10] Alfred Brett, a leading Belfast solicitor, was soon added to the Committee. After his death at the age of ninety he was succeeded by his grandson: together they served continuously for fifty-five years until 1993.

McCausland made a gift of Rough Fort, a thousand-year-old, one-acre ring fort. By the outbreak of war the Committee had also taken on 179 acres of sandy beach and limestone cliffs at White Park Bay on the North Antrim coast. Following the National Trust Act of 1937, the Ulster Parliament agreed to exempt from estate duty property given or devised to the Trust and there was also an understanding that subscriptions from Northern Irish members would be used in the province.

In the same year special arrangements had to be made in the Isle of Man. The Trust was notified that the Calf of Man, the small rocky island just off the coast and an important bird sanctuary, was to be sold. The Trust did not have the £3000 asking price, but their informant bought the island himself and gave it to the Trust. Christopher Gibbs was sent to attend the first meeting of the local Committee, who were displeased to hear that the Trust had appointed a warden without consulting them. Unfortunately the man was not only rude to the Committee but was generally unsatisfactory and eventually had to be sacked. The Calf's management was then handed over to an organization named the Manx Museum and National Trust, which assumed full ownership in 1986.

There was no separate committee or agent for Wales until 1945. Progress there had been slow since the first four acres above Barmouth had been given to the Trust, but by the outbreak of the war the Trust owned several fine stretches of land. These included the Sugar Loaf, dominating Abergavenny and a substantial stretch of the valley of the Usk; Dolmelynllyn, with its spectacular waterfall above Dolgellau;

and some hundreds of acres in the heart of Snowdonia given by Clough Williams-Ellis and two others. In addition, there were a few heterogeneous historic monuments – the Roman fort outside Caernarfon, a medieval house in Conwy, a Tudor merchant's house in Tenby and a tower above Monmouth built as an eighteenth-century gentlemen's club.

In 1939 the Rev. Herbert Lloyd-Johnes gave the Trust the 2500-acre Dolaucothi estate near Lampeter in South Wales; this contained a somewhat marginal gold mine which has been intermittently worked since the time of the Romans. Lloyd-Johnes, whom Matheson found difficult and eccentric, said that he wished the property to be held as a memorial to the Johnes family, which had owned the land since the time of Henry VII and into which Lloyd-Johnes, originally only Lloyd, had married. The Trust accepted the property but must have been a little concerned when they received a letter from the Revd Herbert Lloyd-Johnes' elder son:

> The National Trust should know that this gift of our family property has been made without the consent or approval of my mother, my brother, my sister or myself who are the proper heirs. I myself returned five weeks ago from France having been nearly a year abroad on active service, and this news came as a great shock. I must let you know that about ten years ago my mother obtained a legal separation against my father on grounds of cruelty before a special jury and that all her children supported her. This donation is a case of pure vindictiveness and I feel duty bound to let you know.[11]

After looking into the circumstances, Zetland wrote to the aggrieved son expressing sympathy and giving an assurance that the Trust would care for the estate but saying that there was nothing further he could do.

Money and membership were continuing preoccupations during the 1930s. The Publicity Committee believed that it

Octavia Hill, 1882

Robert Hunter, *c.* 1890

Hardwicke Rawnsley with his son Noel, and Beatrix Potter, 1885

Duke of Westminster, caricature by Ape, 1870

John Bailey, 1926

G. M. Trevelyan and Sir Charles Trevelyan, 1945

Left: Vita Sackville-West and Harold
Nicolson at Sissinghurst, 1938

Below: Evelyn, Duchess of Devonshire,
mending tapestries at Hardwick Hall,
Derbyshire, 1950

Earl of Crawford, Jack Rathbone, Earl De La Warr and Hubert Smith at an Annual General Meeting in the 1950s

Viscount Esher in 1954 with his wife, who accompanied him on most of his visits

James Lees-Milne, 1975

Left: Queen Elizabeth, the Queen Mother, and the Earl of Antrim at Rowallane, Northern Ireland, 1954

Below: Cubby Acland (right) and tenant farmer at Yew Tree Farm, Cumbria, 1977

Above, left: Jack Boles

Above, right: Lord Gibson

Right: Angus Stirling at
Chastleton, Oxfordshire, 1993

Philip Yorke at Erddig,
North Wales

Henry Harpur-Crewe at Calke Abbey,
Derbyshire, 1984

Clergy House, Alfriston, East Sussex, 1894

would be possible to enlist much wider support and to promote the acquisition of more properties if a member of staff
were more often available to speak and, most important, to
make personal contacts. Better-organized publicity, with
dozens of lecture tours and a film together with help from the
press, particularly *The Times*, in publishing announcements
and promoting appeals, at last brought a significant increase in
members. The number rose from 2650 in 1932 to 7100 in 1939
and the growth was consolidated by an annual news bulletin,
which had become a quarterly by the time that the war forced
its abandonment. The Trust's identity was further strengthened by the introduction of the now familiar omega oakleaf
symbol to replace the unsightly noticeboards with their lists of
bylaws. It was designed in 1935 by Joseph Armitage and was
chosen in preference to Eric Gill's lions as being a less com-
monly used symbol and easier to reproduce.

One difficulty was that the Act of 1907 had fixed the membership fee at ten shillings and, although inflationary pressure
was not great, the effects of the First World War meant that
this should have been about £1 if the original value were to be
restored. No change could be made without a new Act of
Parliament and this did not come until 1953. The annual
reports refer repeatedly to the need for more income from
subscriptions, although the 1935–6 report is more encouraging:
'If this improvement can be maintained, the Council will be
encouraged to take more risks in accepting properties and
generally to adopt that more active policy which they, in common with some of their critics, would wish to adopt.'[12]

In September 1939 the Trust was the owner of 58,000 acres
and had another 27,000 acres under covenant. It had raised
and spent half a million pounds on preservation, but its
general endowment and reserve funds stood at only £12,498,
and its membership subscriptions produced only £6173. The
gross income from properties, including income from admission fees and special endowments, came to £27,553. And there

was a considerable question mark over the resilience in war-time conditions of even these limited amounts.

Membership dipped at first but then recovered, so that the Trust made a net gain of 750 members over the war years. In 1945 membership subscriptions amounted to £6600, total income to £14,000, and the general endowment fund stood at £349,000. The unforeseen change was the massive increase in land ownership. The 1939 annual report set out the aims for the duration of the war: 'The Trust's foremost aim in wartime must be to survive the war with its buildings in sound repair, with its farms productive and in good order, with the beauty of its woodlands and open spaces unspoiled and with its financial stability maintained.' It went on: 'the continuous endeavour to promote the acquisition of new properties must cease'.[13] In fact the Trust emerged in 1945 with many of its buildings in disrepair, but owning 112,000 acres, 53,000 acres more than in 1939, nearly all given or left by bequest; in addition some 40,000 acres were protected under covenant, an extra 13,000.

This increase in land came at a time when the government was pressing for every acre in Britain to be ploughed up. The Trust was one of the first amenity societies to voice caution. 'There is a real danger', the 1941 report points out, 'that the government may come to think that land which does not produce certain types of crops is barren and wasted. This would be a foolish view.'[14] The Trust's concern lay in such areas as the Lake District and the Peak District, which were not suitable for crops and would, the Trust argued, benefit the nation much more as light grazing and as places to visit.

The largest gift of land during the war and the largest lifetime gift ever to come to the Trust was the 17,000 acres presented by Sir Richard Acland. Sir Richard was descended from a Devon family, many of whose members had been MPs, and in 1935 he was himself elected as one of the few remaining Liberals. But he did not fit easily into the party framework and in 1942 he joined J. B. Priestley to form the Common Wealth party, advocating common ownership for reasons of ethics as well as economics. When in 1945 he lost his seat he joined the Labour Party, and two years later was elected a Labour MP, resigning in 1955 in protest against the manufacture of the hydrogen bomb. It was entirely in accordance with his principles that in 1943 he should decide to give his substantial estates to the Trust 'on the grounds that service to others, not private gain, must be the mainspring of all human action', leaving himself with very little on which to live.

At the ceremony when both estates were handed over he said, 'The ownership of such an extensive property . . . was an increasingly heavy burden on my conscience.'[15] Another burden which contributed to his decision was the threat of destructive death duties. He continued: 'Death duties today are intended to reduce the size of property holdings on the death of each owner. On my death either Killerton or Holnicote estate would have to be sold and on the death of my son the remaining estate would have to be broken up. If I had continued to own these estates for my own life, then in two generations both would have ceased to exist. Now they are safe forever.' Sir Richard's moral courage and total integrity were expressed in his fine sensitive features, though J. B. Priestley found him 'intolerant, fanatical, tactless and humourless'. (Some people felt that Priestley had his disadvantages too.)

Sir Richard's decision was not greeted with universal approval. In the correspondence columns of *The Times* he was accused of abandoning his duties as a landlord and inheritor of land. Lord Herbert wrote, 'That a man should lightly get rid of his responsibilities and thereby save expenses, is not a matter of congratulation.'[16] G. M. Trevelyan, who had played a part in the negotiations, came to his defence: 'No one with knowledge of the facts could challenge Sir Richard's generosity or could possibly suppose that his gift to the Trust relieves him of heavy burdens.'[17]

The land in question consisted of the two estates of Killerton in Devon and Holnicote in Somerset. Killerton is an agricultural estate seven miles north-east of Exeter which includes the villages of Broadclyst and Budlake and more than 1000 acres of woodland. The house is of no great architectural merit but the arboretum and shrubberies laid out in the late eighteenth century are of rare distinction. The Holnicote estate, which lies on the northern edge of Exmoor, includes the high windswept moorland rising up to Dunkery and Selworthy Beacons leased to the Trust by Sir Richard's great uncle during the First World War, as well as several picturesque villages and Horner Wood with its ancient oak and ash trees.

The stream of smaller gifts and bequests continued unabated during the war and in the year 1940–1 a record number of very diverse properties was presented to the Trust, including Watlington Hill on the Chilterns escarpment (given by Esher), Morden Hall estate on the River Wandle in the suburbs of south-west London and the East Lancashire Stubbins estate of woods and fields leading from the Rossendale Valley up to the moors. In addition, the Trust's land-acquiring initiatives were resumed in the middle of the war despite earlier warnings that such activities would be beyond its wartime resources. Some of these acquisitions were of great importance. In 1943 not only was land purchased in the Lake District but Avebury – with Stonehenge, the most important pre-historic site in the country – was bought with the help of the Pilgrim Trust and others. Next year Flatford Mill in Suffolk, the subject of some of Constable's best-known paintings, was bequeathed to the Trust, and Blaise Hamlet with its picturesque cottages designed by John Nash was acquired with the help of the Council for the Preservation of Ancient Bristol. In the last year of the war the Trust managed to raise £75,000 to buy the 4000-acre Clumber Park in Nottinghamshire. This included a remnant of Sherwood Forest and the three-mile lime avenue leading to the site of the Duke of Newcastle's mansion, which had been demolished shortly before the war. Because of its location within easy reach of Sheffield, Nottingham and several coal-mining districts, eight local authorities made substantial contributions to the purchase price and agreed to pay for subsequent maintenance.

Not all the Trust's acquisitions were equally well chosen. The Trust was given a plot of land in the village of Swanton Morley in Norfolk on the somewhat tenuous ground that it had once been the site of a cottage thought to have been lived in by the ancestors of Abraham Lincoln. But after an impressive ceremony at which Lord Zetland was handed the deeds by the then American Ambassador, John Gilbert Winant, who himself looked rather like Lincoln, it turned out that the ancestors had lived in the Angel Inn. As James Lees-Milne put it, 'the Trust had been left with a useless plot of land, an old chicken run of no beauty and less historic interest which it can never get rid of except by special Act of Parliament'.[18]

Despite the difficulty of recruiting staff during the war, the vast increase in the Trust's holdings made it necessary to begin appointing more qualified staff. In 1942 Hubert Smith was appointed to be the first chief land agent. After Balliol, where he had been a friend of Matheson, and an unsuccessful attempt to set up in farming on his own, Smith had managed Lord Astor's farms at Cliveden and in the Borders. Despite his own lack of professional training, it was he who set up the Trust's regional structure of estate management. He was assisted by Christopher Gibbs, another Oxonian but one who had subsequently qualified as a land agent; Gibbs had joined the Trust before the war and returned in 1943 after a spell in the army. Margaret Sach, who was for a time secretary to both men, recalls Smith as a thinker who wrote very good reports, but was slow and slightly lazy, in contrast to Gibbs, who was quick-thinking and a 'great workhorse'. Gibbs was not unnaturally somewhat put out by the appointment of Smith above him while he was in the army: 'his nose was out of joint and it didn't get in joint until Hubert retired [in 1961] and he couldn't get in the room quick enough'.[19] Nonetheless he put up with Smith's quirks and tantrums uncomplainingly.

At the end of the war in Europe the Trust was fifty years old and was ceasing to be 'a dedicated and amateur group, quixotic and sometimes haphazard'.[20] It had become a land-owner on a scale that many would have thought inconceivable even in 1939 (though it was much smaller than the Forestry Commission) and it was beginning to introduce a professional management. Besides open spaces, small amenity woods, archaeological sites and buildings of every period and size it owned large agricultural estates and forests which had to be properly husbanded. Some 800 cottages and well over 100 farms gave it a responsibility for rural housing and agricultural policy. Its extensive holdings on the coast, in the Lake District and amid other outstanding landscapes brought it in touch with problems of national planning and the National Parks.

But in a lengthy memorandum written towards the end of 1944 Matheson pointed to some of the Trust's weaknesses: the concentration on expansion had meant that funds for proper upkeep had been strained; most of the farms and cottages had been acquired recently and only a small proportion had been brought up to even a minimum standard of repair; and some country houses had been taken on with little endowment and without contents. He noted a widely held feeling that the Trust had been ineffective in managing nature reserves. Looking to the future he foresaw financial problems, exacerbated by the general rise in wages and the need for a larger staff in order to improve maintenance and to cope with more visitors.[21]

The Trust, like many institutions and individuals had, since the hinge year of 1942 been endeavouring to peer ahead to the post-war world. The expectation was that it would continue to receive country houses, but that increases in acreage would be less spectacular and that heavy taxation would make it difficult to raise large sums by subscription. But would the Trust have a lesser role in the areas to be covered by National Parks? Would a new agency take over the preservation of nature reserves? What would be the effect of comprehensive land-use planning? And what would be the Trust's relationship to the state?

CHAPTER SIX

Peace and
Socialist Utopia

O N 26 JULY 1945 the Labour Party, against most expectations, was elected to power. Clement Attlee became Prime Minister of the first majority Labour government and proceeded to carry out the most far-reaching reform programme of the century. During the first three years the legislative framework of the welfare state was put in place; the public utilities and the coal industry were taken into public ownership; a town and country planning system was enacted and a series of new towns initiated (with Clough Williams-Ellis as chairman of the first); and India and Pakistan became independent states. All this took place while millions of men and women were being demobilized and industry was being switched from arms to civilian goods and while the government was staggering under a deficit of foreign exchange no longer concealed by American Lease–Lend and exacerbated by the wartime sale of foreign investments. For most families these difficulties showed themselves in continuing shortages of food, fuel and household goods. For the Trust they meant building permits, petrol rationing and continued requisitioning.

The Trust was faced with a new political map which it was not particularly well equipped to read. Although it had always sought support from all political parties its leaders did not enjoy the close personal relationships with Labour ministers that they had with the Conservatives. Whereas Zetland had been a Conservative MP and had served in the government until 1940, the only member of the Executive Committee to be on terms of personal friendship with any of the leading members of the Labour government was Chorley, who had been on the staff of the London School of Economics with Hugh Dalton, the Chancellor of the Exchequer, and became an Attlee-created peer. Harold Nicolson was soon to join the Labour Party and fight an unsuccessful by-election on its behalf, but in 1945 he had stood against Labour and his wife was Vita Sackville-West, cousin of Lord Sackville, the donor of Knole in 1946, whose social world was more to Nicolson's taste than that of the local Labour club. Zetland had not followed up suggestions made by Attlee for appointing a few members more in touch with the trade unions and Labour, although the names put forward included such eminent figures as Professor R. H. Tawney, the historian, A. D. Lindsay, the Master of Balliol, and Sir Walter Citrine, the general secretary of the Trades Union Congress, as well as Arthur Creech Jones, a Labour MP, who had introduced the Access to Mountains Act in 1939 and was to be Secretary of State for the Colonies.

In the month the war in Europe came to an end, the Trust's head office was moved from Buckingham Palace Gardens to new premises in Queen Anne's Gate, Westminster, in which street it has since remained (although in 1983 it moved a few yards to larger offices). The gift of two early-eighteenth-century houses came just at the time when the extra staff could no longer be crammed into the existing building.

In the same year new officers took over several of the most important positions in the Trust – a new chairman, treasurer and secretary, following a new president the previous year when Queen Mary had agreed to continue the royal connection and fill the vacancy left by the death of Princess Louise. Zetland had first asked the Earl of Harewood whether his wife, the Princess Royal, would be interested, but had received a sharp rebuff:

I am a very bad person to approach regarding the activities of the National Trust as I seem to be the only person in England who disapproves of it. First, as an attempt to save country houses for the benefit of the descendants of the owner . . . I cannot see that it is any help at all. Secondly, a more serious consideration arises from the course now being followed by the National Trust. You are becoming very large owners of land, and I find it difficult to approve this procedure without facing the implication that nationalization . . . is favoured by you. . . . Perhaps you can give the Princess Royal a sound answer to this latter point, and assure her that she cannot be accused of connecting herself with an organization which implies the approval of land nationalization, and in that case I think she would be inclined to be your president.

Zetland apologized for his 'faux pas' and decided to aim rather higher and approach Queen Mary instead. This time he received an enthusiastic response: 'The Queen is keenly interested in the work of this body and Her Majesty would be really glad to have this personal association with its aims and achievements.'[1]

Zetland's departure did not come as a surprise. He was sixty-nine and had served as chairman for thirteen years. He may, as Gibbs later recalled, have had 'no knowledge or interest in natural beauty and have known no properties', but he read the committee papers with care and these gave him a grasp of the Trust's work. He took a real interest in relations with the government and in the organization of the Trust, supporting the appointment of a full-time lawyer to the staff, and on one occasion noticing that two words had been omitted from a memorandum defining the duties of Lees-Milne. Norman was again offered the chair and again declined. This time he felt that he was too old and that someone younger was needed. Again Norman found a suitable successor – David Lindsay, the twenty-eighth Earl of Crawford. He became the

Trust's longest-serving chairman, and during his twenty years he presided over its transformation from a small and little-known society to an institution of international standing.

It is at first sight unlikely that Scotland's premier Earl should head a charity devoted to saving the countryside in England, Wales and Northern Ireland. However, Scottish leadership is not unusual in England and he had inherited a fine early-ninteenth-century house and estate on the edge of Wigan in Lancashire. He was peculiarly well suited for his task. Not only was he brought up in 'the circles in which it is so important for the chairman of the National Trust to be at home',[2] but he had been introduced to all aspects of the amenity movement by his father, who had been the first president of the CPRE as well as chairman or trustee of several major arts institutions.

David Crawford was born in 1900 and enjoyed the conventional upbringing of a young man from his background – Eton, Oxford and then a period as honorary attaché to the Rome Embassy before becoming Conservative MP for the Lonsdale Division of Lancashire from 1924 until he succeeded to the earldom in 1940. He acquired a wide-ranging knowledge of art and books from the family collection of paintings and the famous *Bibliotheca Lindesiana*, most of which he was obliged, much to his distress, to sell together with his Wigan mansion in order to meet estate duty. His talents and training made him an ideal chairman of the National Gallery, the Royal Fine Art Commission, the National Art-Collections Fund and the National Gallery and Library of Scotland. He was also a long-standing trustee of the British Museum. Holding some of these offices for twenty or thirty years gave him an exceptional position of sustained influence. 'We have no one who can speak with your authority, no one else who carries weight beyond the confines of our world,' wrote Robin Fedden, the Historic Buildings secretary.[3]

As a chairman of committees he did not, in Chorley's view,

match Zetland, perhaps because he was too kind to cut short even the most long-winded member. But he usually succeeded in obtaining a consensus in line with his own views, and his consideration endeared him to younger members of the staff as well as to his older colleagues. Among the latter was Ruth Dalton, wife of the Chancellor and not naturally one of his admirers. 'I cannot imagine the Trust without your charm, your wise leadership, immense knowledge and psychological understanding,'[4] she wrote on his retirement. Chorley also recalled his charm – it was 'not of the type which cloys, [he] put you at your ease by making you feel that your opinion was worthwhile' – and admired his industry, never sparing himself whatever task he undertook.[5] Perhaps Crawford's greatest contribution to the Trust was to set standards for managing its houses and collections during the years when some of the most important properties had to be repaired and reordered. 'I wonder if you realize the supreme importance that this automatic sense of aesthetics in you has had for the Trust,'[6] commented Lord Rosse. His wonderful eye for buildings, gardens and countryside, as for pictures, was appreciated by both land agents and aesthetes. 'To travel in his company was to feel as though one's sight had suddenly been restored after a lifetime's blindness,'[7] wrote John Gaze, later the Trust's chief agent.

Crawford was at first doubtful about accepting Zetland's offer of the chairmanship. 'Already I have more work than I can properly carry out. . . . I live in the country far from London. . . . I only come to London once a month.'[8] But Zetland persuaded him that he would not find it too onerous. 'The main task is that of presiding at the meetings of the Executive Committee which meets once a month, and attending meetings of the Finance and General Purposes Committee. . . . Apart from these specific tasks . . . general supervision over the work of the Trust . . . [and] a fair amount of correspondence . . . [the] only other call in normal times is attendance at

126

a certain number of formal functions.' One member of the Council told him that the Trust needed a good deal of pulling together but the Committee 'seem to be nice people (apart from the secretary who is a very difficult creature)'.[9] Eventually Crawford agreed to take on the chairmanship provided that meetings were moved to the second Friday of the month in order to fit in with his other London engagements (though this meant that Trevelyan could never attend during term). He restricted himself to the duties Zetland had mentioned and, although he practically never used the telephone, he sent a stream of letters to his various institutions, never less than a dozen a day, expressing his thoughts concisely in his semi-shorthand hieroglyphics.

But he was not easy to get hold of and difficulties occasionally arose from his firmly held rule of not leaving Fife more than once a month. There could be delays in meeting ministers (in 1946 it took four months to find a day to see Lewis Silkin, Minister of Town and Country Planning, who had complained that he was being cold-shouldered by the Trust). There were requests to Esher to settle matters as it could take a long time to get a reply from Crawford. And, most important, when personal differences arose between the secretary and a member of the Executive Committee, they were more difficult to sort out by correspondence than they would have been face to face.

Crawford's second-in-command was Esher, the powerful chairman of the Finance and General Purposes Committee, who had effectively run the Trust during the war and might reasonably have expected to succeed Zetland. He would probably have been an excellent choice for he knew the properties well and, as an obituary pointed out, 'his finesse in the handling of committee work and his long experience in the niceties of compromise and evasion which it entails gave a professional smoothness to his performance . . . which was all the more effective for the fact that he was, for the most part, and in the best sense of the word, an amateur'.[10] But he was sixty-five and his wit could cause offence. Trevelyan stayed on as chairman of the Estates Committee, which had not functioned during the war, and the Publicity Committee was relaunched and was soon to have an energetic chairman in Edward Keeling MP. Cecil Lubbock, the Trust's honorary treasurer, was replaced by Edward (Ruby) Holland-Martin, a director of the Bank of England and honorary treasurer and vice-president of the CPRE. 'I am sure you could not get a better man,' wrote Crawford to Esher. 'I have known him for many years and am very fond of him. I am sorry to hear of Lubbock's resignation but I expect that Lubbock is right in suggesting he is less in touch with the City than Ruby, even though the latter no doubt will soon be nationalized.'[11] Holland-Martin was much respected. He was assiduous in visiting properties and entered into the spirit of what was being done, but he could be distant and cold: 'as friendly as I am sure a swordfish can be', remarked Lees-Milne.[12]

Crawford was anxious to bring new blood on to the principal committees: 'though Pleasure may be at our Helm, Youth clearly is not at the Prow', he commented. By 1950 he had enlarged the Executive Committee to twenty-three, bringing on thirteen new and younger members, of whom seven had hereditary titles. They included Earl de la Warr, who became chairman of the Estates Committee, the Earl of Rosse, who was to follow Esher as chairman of the Historic Buildings Committee, the Earl of Antrim, who was chairman of the Committee for Northern Ireland and was to succeed Crawford as chairman of the Trust, and the Hon. Harold Nicolson, who became vice-chairman.

At the prompting of Trevelyan, Chorley was also added to both committees. 'I should like to get Prof. Chorley on to the Committee. He is slightly different to us, more in touch with the democratic world. He is most loyal to the Trust and has often given good advice from experience to which he is closer than we are.'[13] Chorley himself wrote to Crawford suggesting that it would be helpful to have more support from the Labour benches on the Executive Committee, 'but I hope that you will not think I am trying to form a "bloc"'.[14] He suggested Walter Scott-Elliott, Member for Accrington, an Etonian Scottish hill farmer who was a rather detached Labour MP for a few years after 1945. In reply Crawford agreed: 'On paper we *look* a too "conservative" body and in any case political strengthening in the House of Commons on the government side seems most desirable. Are you sure of your man?'[15] There were already two Conservative Members on the Executive Committee and from that time there has always been at least one Member from each of the two main parties. Scott-Elliott served for twenty-one years (from 1946 to 1967) without ever taking a very active part, but Ruth Dalton, who was appointed the same year, became a tower of strength. She was able and sensible, and had time to visit properties in different parts of the country. Despite her rather forbidding appearance, she was very kind to younger and more vulnerable members of the staff.

Finding a new secretary proved the most difficult problem. Esher had clear views on the subject: 'The secretary's business is to drive this team of temperamental experts, to be intelligent about the subjects with which they deal, to be of sufficient weight and character to deal with government departments and important people of all sorts.'[16] The Trust appointed three secretaries between 1945 and 1949 before they found a man even remotely meeting Esher's criteria and had it not been for him, Crawford and Norman, assisted by Lees-Milne and Anthony Martineau, the Trust would barely have stayed afloat during these years.

Martineau was an able, meticulous solicitor of some private means, who, when he had been invalided out of the army, had written to enquire whether there was a vacancy with the Trust. Matheson appointed him as assistant secretary with the intention that he should become the Trust's first full-time lawyer, taking over from Messrs Horne & Birkett, the solicitors who had hitherto managed the Trust's affairs. He was a man of high principles, familiarly known as the 'archangel', who used to retreat for private prayer each evening, even when he had guests to dinner. His principles caused him to threaten resignation whenever he disapproved of what he felt to be unduly favourable terms for donor families insisted on by Esher. He was formally appointed legal adviser in May 1945. The growth in the Trust's legal work made the change necessary, but marked the end of an era for Benjamin Horne:

> I attended the meeting of the Committee when the three founders discussed whether or not the offer of the Barmouth Cliff land should be accepted and I have been at most of the meetings during the last fifty years and have myself done or been responsible for practically all the legal work ever since, as over 1500 bundles testify. It has been my main interest in life . . . and throughout a considerable allowance has been made of my firm's charges.[17]

George Mallaby was the first of the three secretaries to be appointed. He was forty-three years old and had been a well-liked headmaster of St Bees School, near Whitehaven, before becoming special commissioner with the aim of bringing new industries into the depressed area of West Cumberland. Since 1942 he had worked in the Cabinet Office. He served at the Trust for barely a year before accepting a post in the Ministry of Defence. The reason he gave to the Trust was that this was a bigger job and one for which he was better qualified. But he told his family that at the Trust he was not allowed enough responsibility or freedom to take decisions. He had already asserted his authority, introducing more method, despite difficulties with Lees-Milne and Hubert Smith, whom he found hopelessly unbusinesslike. His departure in the spring of 1946 brought a typical response from Esher: 'What a nuisance – but I think inevitable as I always felt that Mallaby did not feel happy in the free air of the Trust and pined for the Civil Service atmosphere.'[18] Crawford agreed, doubting whether Mallaby had the judgment required for the job.

On the second occasion the Civil Service and the Oxford Appointments Board were consulted and other names were canvassed, including that of Harold Nicolson. In his diary entry for 10 May 1946 Lees-Milne notes, 'Called on Harold Nicolson at six. He said he would not consider the NT secretaryship because he was intent on returning to politics. But advised that the status should be elevated to director-general or some such term and the salary raised.'[19]

Nicolson's advice was not taken until 1968. The salary of £1250, only £350 more than Matheson had received in 1939, was not sufficient to attract high-powered candidates. Moreover, the status of the post was unsatisfactory, it being unclear whether the chief agent was responsible to the secretary or on a par with him. In June 1946, Vice-Admiral Oliver Bevir, aged fifty-five, was appointed in preference to John Arkell, aged thirty-seven, who had been assistant secretary of the CPRE, had reached the rank of major during the war and went on to a senior position in the BBC. Arkell was later to produce an influential report on the Trust's relations with the public and its members and would undoubtedly have been a better choice.

Bevir, a typical sailor with a quarterdeck manner, found the work difficult and exhausting. At his interview he had said that he was looking for something to do before retiring, but this danger signal was ignored. Lees-Milne described the Admiral's predicament: 'Our new secretary Admiral Bevir in attendance although George Mallaby was in the office for the last time. Admiral seems jolly, if too inclined to be facetious. Of course he has been warned by the chairman that we are not a bureaucratic team of experts, but a dedicated group of happy-go-lucky enthusiasts, who ought not to be bossed about.'[20] And eleven days later: 'Today we had a Reports Committee at which the bewildered Admiral was present. He does not know much about the National Trust and is, besides, not an intelligent man. He pierces one with a cold blue eye like a schoolmaster trying desperately to assess a new boy; only he is the new boy.'[21]

Little over two years later Bevir, to Crawford's irritation, announced that he wished to retire and the search for a secretary recommenced. Jo Grimond, the secretary of the National Trust for Scotland and a future leader of the Liberal Party, was canvassed, but his eye was already on the House of Commons. The successful candidate was J. F. W. Rathbone. He had been suggested by Eardley Knollys, who commended his organizing power and his ability to get on with all kinds of people, adding in a postscript, 'He is NOT a Philistine, though his aesthetic taste is limited and undeveloped.'[22] Esher reported that the selection committee had some good men from whom to choose, but were all agreed on Rathbone, a choice welcomed by Lees-Milne, who had known him at Oxford.

Jack Rathbone was born in 1909 of a Liverpool family with a record of public service. He was educated at Marlborough and New College, Oxford, before going on to qualify as a solicitor. During the war he became a colonel and was then appointed Director of the Ministry of Justice in the Control Commission for Germany. Rathbone was secretary for nineteen years. Unmarried, he was a short, dapper man with great charm and the ability to get on with suspicious Labour ministers as well as with awkward donors. With his chairman in Scotland, it was he who forged the scattered, hard-pressed agents into a team, inspired them with his own enthusiasm and, on a shoestring, led them through a period of very rapid growth. But he did not exert a strong leadership – John Gaze recalls his wavering on an important long-standing dispute about expenditure[23] – and he changed his views chameleon-like to fall in with those of others. Always anxious not to hurt other people's feelings, his gushing manner in conversation and correspondence did not earn respect. Writing two years after his appointment Crawford reported after a minor disagreement that 'The row (as rows do) reduced Rathbone to hysteria.'[24] At the time of his retirement Mark Norman, a long-standing member of the Finance Committee, summed him up as 'excellent for a small National Trust. Intelligent, passionate, mercurial, disorganized and totally devoted to the Trust.'[25]

Rathbone inherited an organization that had been substantially strengthened since the end of the war. As early as 1941 Matheson had taken the view that the Trust's properties had become too numerous and diverse to be managed by unpaid local committees assisted by local firms of agents. Apart from the need to ensure that endowment land yielded the anticipated income, Matheson pointed out that 'the work calls for a realization of the Trust's special responsibilities, good taste and a keener appreciation of natural beauties than is common among many land agents'.[26] His solution was to appoint a number of full-time area agents, responsible to the chief agent in London. By June 1946, nine area agents were in post. They had to cover extensive areas without much help and to assume responsibilities well beyond the normal running of existing properties. The idea was that they should become the driving force for all the Trust's activities within their remit, should co-operate with local committees and establish good relations with landowners and local authorities, and should devise some means of efficient management without losing the human and personal touch. Bruce Thompson had proved in the north of England that these ambitious terms of reference could be carried out. Several of the new agents were to make their mark on the Trust during the next twenty or thirty years. Most noteworthy was Cuthbert (Cubby) Acland, Sir Richard Acland's younger brother, who moved to the Lake District in 1949 and became known as 'Cock of the North' during his twenty-four-year reign. A red-faced, beaky-nosed man pulsating with energy, as incisive in manner as in mind, he set the highest standards for managing farms and woodlands and did much to build up the Trust's estate. By the late 1940s the area agents were sufficiently well established to take responsibility for day-to-day business and, though they felt that not enough decisions were delegated to them, the Estates Committee was able to meet less frequently and to concentrate on matters of policy. This suited De La Warr, its new and popular chairman. No highbrow, unlike his predecessor Trevelyan, he was well versed in estate management and got on well with the agents. Trevelyan's retirement marked the separation of the leaders of the Trust from the access and conservation movements in which the Trust had played a significant role since its founders' time.

Born in 1900, De La Warr was educated at Eton and Oxford, but instead of joining the Conservative Party like most of his friends he served as a parliamentary secretary in the 1929–31 Labour government, was a member of the National Government until 1940 and returned to office as Postmaster-General in the Conservative Government of 1951–5. Although very much one of the establishment (he was chairman of the Country Landowners' Association), he was unusually open-minded and was one of the first Committee members to recognize the demands of the visiting public for signposting and other facilities which are now taken for granted.

Only one or two of the agents had much feel for aesthetic or architectural matters and so in 1947 it was decided to introduce 'representatives' with the taste and experience to deal with these aspects of historic buildings. The proposal had been aired by Mallaby in December 1946 and its implementation is described by Lees-Milne: 'Lord E[sher] has approved Michael [Lord Rosse]'s memorandum on the proposal that we should have area representatives as well as agents. Lord E wants to call them area artists, but Lady E and I think that a bad name, which will cause resentment in those who are not artists. I have engineered this little scheme.'[27]

Full-time representatives, including Eardley Knollys, were

appointed in three areas, and elsewhere honorary representatives came forward to act in a voluntary capacity, among them George Howard, the owner of Castle Howard, one of the largest private houses in the country. At this stage the representatives were not knitted into the area organization, lacking even a desk in the area office and having no regular contact with the area agent; as a result there developed a schism between the agents (known as 'mangelwurzels') and the representatives (the 'lilies of the field'), with divisive consequences for the future.

The representatives had to take responsibility for an enormous range of pictures, furniture and objects of art, some of great value. To the historic collections associated with particular houses were added Lord Bearsted's more recent collection of old masters, porcelain and furniture at Upton House in Warwickshire and Anthony de Rothschild's collection at Wing in Buckinghamshire. In 1948 Anthony Blunt, Keeper of the King's Pictures and director of the Courtauld Institute (later to achieve a greater but less desirable fame), was appointed honorary picture adviser. Two years later the first attempt was made to look seriously at important libraries, which until then had been treated as little more than typical country-house furnishings. Robert Gathorne-Hardy was brought in to examine the books at Blickling and Charlecote.

The Trust viewed the Labour government with a mixture of hope and suspicion – hope that National Parks would at last be designated, and suspicion that the Trust might be placed under state control or even nationalized. What they did not expect was that Hugh Dalton, the egalitarian Chancellor of the Exchequer (in his first two Budgets he raised death duties to 75 per cent and put up surtax), would perform 'one of the most forward-looking and imaginative acts of government in this century in relation to general culture'.[28] Dalton was an enthusiastic fell-walker and a supporter of the Trust. Five months after taking office he sent instructions to his

permanent secretary, Sir Edward Bridges, also a supporter of
the Trust and later a member of its Executive Committee, to
work on a proposal for:

> Payment of all proceeds of sale of War stores into a special
> National Estate Fund – to be spent on real estate only, e.g.,
>> in providing National Parks,
>> in aiding National Trust,
>> in making good Death Duty Revenue when payment of
>> this is in land.[29]

In his spring Budget of 1946 Dalton resurrected a provision
in one of Lloyd George's Finance Acts which permitted land to
be handed over in payment of death duties, and set aside £50
million for a National Land Fund to reimburse the Inland
Revenue. In his speech he said that the Trust might be asked to
hold such land and might also receive land which the Fund
would have power to buy. He emphasized that no money
would be paid to the Trust, but a few weeks later he offered to
pay the Trust, on a pound-for-pound basis, a sum equal to the
money raised by the Trust's Jubilee Appeal up to a maximum
of £60,000, recognizing that there would be expenses involved
in managing the properties transferred through the Land
Fund. Dalton intended the Fund to be a war memorial. 'I
should like to think that through this Fund', he told the House
of Commons when announcing its inception, 'we shall dedicate some of the loveliest parts of this land to the memory of
those who died in order that we might live in freedom, those
who for our sake went down to the dark river, for whom
already the trumpets have sounded on the other side.'[30]

Writing to Crawford, Dalton pointed out that handing over
land to the Trust in this way was something of a novelty. He
added:

> As you know, I am a most empirical socialist. I once des-
> cribed the Trust in a book I wrote as a typically British
> 'example of Practical Socialism in action'. I added, 'It has

behind it a fine record of public service and commands a widespread public goodwill. A Labour government should give it every encouragement greatly to extend its activities.' This is all I'm trying to do.[31]

The Land Fund was not as successful as Dalton had hoped, though Sir Stafford Cripps, his successor as Chancellor of the Exchequer, enlarged its scope by exempting from death duties land provided as an endowment. Cripps had long been anxious to protect the countryside. In an introduction to a pre-war book of essays edited by Clough Williams-Ellis he had written, 'We cannot go back, we do not want to go back, to the conditions of feudalism, but we must somehow wrest our beauty of the country from the grip of the Beast of industrialism.'[32] In March 1948 he told Esher: 'You are taking a long time to spend the £50 million we have put aside for the National Trust. . . . you can spend as much of the reserve as you like so long as it is spent on the National Trust and not on the precious owners.'[33]

Dalton had expected the Land Fund to be spent over a period of four or five years and when exhausted to be replenished, but by mid-1951 at the time of the fall of the Labour government it had disbursed less than half a million pounds. Fourteen properties had come to the Trust including two houses – Cotehele in Cornwall and Rainham Hall in Essex – as well as Claremont Landscape Garden in Surrey, and some farms, woodlands and coastal land.

The following year the Fund paid out the largest sum so far – £207,000 for Penrhyn Castle in North Wales and 40,000 acres intended for its endowment. But even the largest stately houses and the most extensive estates did not compare in value with works of art. It was not until two Conservative Chancellors of the Exchequer responded to Crawford's pleas for the Fund to take in works of art that payments began to increase significantly. In 1953 R. A. Butler extended its scope to take in objects of historic or artistic interest associated with buildings belonging to the Trust, thus facilitating the transfer of the contents of Petworth at a cost of £553,000; and in 1956 Harold Macmillan included pre-eminent works of art irrespective of their connection with a particular building, which led almost at once to £1 million being paid for art treasures from Chatsworth. But only a year later Peter Thorneycroft, Macmillan's successor as Chancellor, reduced the Fund's capital, which had then reached £60 million, to £10 million. By this date the Fund had paid out less than £2 million; when in 1980 it was wound up a total of just over £19 million had been spent.

Why was the Land Fund not more used, especially in its early years? One reason was that owners were deterred by the difference between the probate valuation used to calculate the sum payable to them and the market value. A second and more important reason was that the Fund was not permitted to make capital payments to the Trust for the endowment of historic houses, a constraint not applied in Northern Ireland when the Ulster Land Fund was set up. Nor were objects of art included until the 1950s. A third reason was that the Fund was not vested in a separate agency with its own trustees and freedom to take its own decisions as was the National Heritage Memorial Fund which replaced it.

The second legislative change introduced by the Labour government which had a direct impact on the Trust was the Town and Country Planning Act of 1947, which, unlike the Land Fund, had been foreshadowed in the party's election manifesto. Based upon three reports published during the war – the Barlow Report on the Distribution of the Industrial Population, the Uthwatt Report on Compensation and Betterment and the Scott Report on Land Utilization in Rural Areas – it was the second important measure brought forward by Lewis Silkin, the Minister of Town and Country Planning, and followed the New Towns Act of the previous year. The Bill introduced statutory land-use planning and nationalized development rights. In essence the system still survives, apart from the financial provisions, which were repealed, although there was some considerable weakening of the controls over agricultural land in the 1980s. It has gone some way towards containing the big cities, protecting the Green Belts and restricting development in the countryside. Nonetheless, land in urban use in England and Wales increased from 6.4 per cent in 1947 to 8.8 per cent in 1969 and 10.2 per cent in 1990, and development has continued to scar the coast and other scenic areas as well as the wider countryside.

The 1947–8 annual report noted that the Act 'is so far-reaching a measure that there can be scarcely an individual, let alone a large land-owning body like the National Trust, which it does not closely affect'.[34] The Trust was not happy with the Bill as it first appeared but, thanks to Esher's advocacy in the House of Lords, the Lord Chancellor agreed to introduce a new clause providing that the Trust's inalienable land would be virtually excluded from the compulsory acquisition powers. Great hopes were placed on the Act, so much so that Chorley wrote seeking Dalton's advice as to whether it would be sensible for the Trust to concentrate on the management of its existing properties and leave the protection of the countryside to the local authorities working under the Act. Less starry-eyed, Dalton advised, 'If the property is really worthwhile always get it for the Trust and make it inalienable. It then becomes completely safe. Silkin has produced an admirable Bill but it does not give anything like such good protection as this.'[35]

The National Parks and Access to the Countryside Act was Silkin's third major piece of legislation. National Parks, advocated by the Trust since its early years, had been recommended by a committee set up by the minority Labour government of 1929–31 and again in a report by John Dower (the son-in-law of Sir Charles Trevelyan) towards the end of the war. Dower defined what a National Park would mean in a country which lacked the vast wilderness areas of North America:

> It would be an extensive area of beautiful and relatively wild country in which, for the nation's benefit and by appropriate national decision and action (a) the characteristic landscape beauty is strictly preserved; (b) access and facilities for public open-air enjoyment are amply provided; (c) wildlife and buildings and places of architectural and historic interest are suitably protected while (d) established farming use is effectively maintained.[36]

Dower's report was followed up by a more detailed one from a committee chaired by Sir Arthur Hobhouse and including Theo Chorley which recommended the establishment of twelve Parks, each to be administered by a committee of which half the members and the chairman would be appointed by the proposed National Parks Commission and the other half by the local authorities. The administrative and access provisions of Silkin's measure left much more power to the local authorities than the Ramblers' Association, of which Dalton had become president, would have liked. Even so, Dalton claimed that but for his intervention the Bill would not have been 'half as good'. He certainly wanted to go further on compulsory land acquisition. 'We shall have no peace around the Peak until we have paid off the Dukes,' he remarked.

The Trust was upset to find that again the clauses relating to compulsory acquisition applied to its inalienable land. But the tactics adopted to try and get the Bill amended were, as Esher pointed out, not very clever:

> I do not think we have been very skilful in our conduct of this political issue. If you merely move amendments in Committee on a government bill, and if those amendments are moved by two eminent Opposition MPs, the government Whips are put on, and nobody dares vote against the government. I do not think we should repeat this procedure in the

House of Lords. It is true that there we might defeat the government, but the House of Commons would not accept that defeat. In my view we should have negotiated privately with Mr Silkin. Now, after 'able speeches' and lost divisions, it is probably too late.[37]

The Trust's cause was not helped by the fact that the two MPs in question, Edward Keeling and Harry Strauss, both of whom were members of the Executive Committee, were unpopular with Labour Members, who felt that they would not discuss any issue except on party lines. Fortunately Cubby Acland had an opportunity to meet Silkin in the Lake District, on the occasion of one of Dalton's well-publicized walking tours with a group of young Labour MPs (not that Silkin, a very urban solicitor, joined the walk). But Acland did not make much headway, Silkin complaining that he had wasted hours in committee over the Trust's amendments and the more he heard about them the more he thought he was right. Rathbone was more successful a week later when Ruth Dalton organized a brief and apparently casual encounter between him and Hugh Dalton: Dalton promised to try and persuade Silkin to put down an amendment at Report stage and encouraged Esher to see Silkin himself, remarking that 'Lord Esher is famous for his wonderful manner with Ministers.'[38] Esher took the advice and Silkin agreed to some, if not all, of the changes sought by the Trust.

Dalton became Minister of Town and Country Planning after the 1950 election and proceeded to implement the National Parks Act as rapidly as the complicated procedure allowed. He designated the first three Parks – the Lakes, the Peak and Snowdonia – though in Chorley's opinion he gave away too much to the local authorities. Always keen to promote able young men, it must have amused him to appoint Cubby Acland to the Lake District Board in preference to Sir Robert Ewbank, chairman of the Trust's Advisory Committee, who thought he had a better claim. Dalton also recommended Desmond Donnelly, 'a live young Labour MP', for the 'rather too old and respectable Executive'.

The Trust feared that the creation of National Parks would affect it unfavourably. Dower had forecast that 'the mere fact that an area had become or was generally expected to become a National Park, would make it difficult for the Trust to raise appeal funds for acquiring new properties therein,'[39] and expected that the Trust's responsibilities in those areas would steadily fade. Matheson concurred and Mallaby took an even more pessimistic view. In a memorandum to the Executive Committee he posed the question whether the Trust would enjoy any 'substantial and active life' after the introduction of National Parks, concluding that the public would come to think it 'quite superfluous to support the National Trust any longer' and 'there would be no national support for Trust acquisitions even in areas outside the Parks'.[40] He proposed that the Trust should become a commission of government, managing not only the National Parks, but taking over historic buildings and ancient monuments from the Ministry of Works. Esher's reaction, which reflected that of the Executive Committee, is recorded by Lees-Milne:

> Esher spoke most eloquently about the Civil Service mind. . . . He began by saying he never believed in hara-kiri. Many people after 1911 thought the House of Lords was doomed and gave up the struggle, whereas it still survives. He believes it fills today another but no less necessary function than it did in 1910. The same could be said of the Monarchy now that it is constitutional. This is why he believes the National Trust will survive, and profoundly disagrees with Mallaby's, the new secretary's, memo on the National Parks question. It advocates defeatism in assuming that the government are bound eventually to take over the Trust's activities.[41]

The Trust's fears proved unfounded. All except two of the areas proposed by Hobhouse have been designated as National Parks, but development has not been fully controlled and access has not been opened up to the extent demanded by the ramblers. Partly for these reasons, Trust acquisitions within the National Parks, far from coming to a halt, increased much more quickly in the forty years after 1950 than they had in the previous forty. By 1990 the Trust owned 25 per cent of the Lake District, compared with about 10 per cent in 1950, and 12 per cent of the Peak District, compared with a fraction of that in 1950.

The Trust was equally pessimistic about the likely effect of the Land Fund on offers of country houses. Nicolson explained why:

> I think that the Dalton Act will alter the situation more than is generally realized. Hitherto owners have felt that in transferring their properties to the Trust they are transferring them not to a government department but to an organization directed by fellow impoverished aristocrats. They will now begin to feel that as the Treasury is going to bag their properties in any case as payment of death duties and as their heirs will never in any case be able to maintain the houses, it would be preferable for them to swallow the houses and to retain such investment capital for their heirs to go to Eton and themselves to go to Barbados.[42]

But Nicolson did not appreciate the effect of Dalton's tax measures on the financial position of historic-house owners. By 1947–8 few people had much more than £5000 a year to spend after tax and only seventy individuals were nominally left with more than £6000; many great houses needed at least £5000 a year each, and some as much as £10,000, to meet the minimum costs of upkeep and essential repairs. The tax relief allowed on 'maintenance claims' was quite insufficient to bridge the gap. There were also serious practical difficulties in reopening a large house in the post-war years when there were persistent shortages. Country houses had been requisitioned for use as hospitals or schools or for the services, and had often been left with considerable damage. Lees-Milne describes Van Dycks used as dartboards, balustrades consumed for firewood and landscaped parks turned into army camps. Furniture had been stored away and if the families had not moved out they retreated to a few rooms which became increasingly difficult to manage as the servants left for the army or the factories. Demobilized footmen and maids had become accustomed to a very different way of life and few wished to return to domestic service. At Ham, for example, there were two gardeners instead of the pre-war twelve, and at Petworth only a house-keeper and housemaid looked after the vast mansion. The 1.3 million people employed in domestic service in Britain in 1931 declined to 250,000 in 1951 and 100,000 in 1961. The situation was summed up by P. G. Wodehouse, writing to a friend in December 1945: 'I wish I could get the glimmering of an idea for a novel. . . . What the devil does one write about these days, if one is a specialist in country houses and butlers, both of which have ceased to exist?'[43]

CHAPTER SEVEN

White Elephants

A RECORD number of country houses passed to the Trust in the five years after the war, including some of the oldest and most renowned in the country. Esher and Lees-Milne no longer expressed disappointment that the Trust was receiving only the second-rate. In 1946 Knole in Kent, with its legendary 365 rooms, fifty-two staircases and seven courtyards, and some of its family portraits and rare furniture, was presented by the fourth Lord Sackville, whose ancestor had been given the house by Queen Elizabeth I. The negotiations had lasted eight years. This was partly because of the need to break the entail for any sons who might be born to Lord Sackville's son Eddy Sackville-West (perhaps an unlikely contingency and one which indeed never happened), before the Court of Chancery would make an order vesting Knole in the Trust subject to a lease in favour of Lord Sackville. Even Vita Sackville-West, who had been brought up at Knole and was obsessed by her memories, recognized the inevitability. 'Of course I cannot pretend I like it but I know that it is the only thing to be done.'[1] When Lees-Milne asked her to write the guidebook she agreed if he would check that the furniture and pictures were still in the same positions as they had been in 1928, when she had last seen the house after her father's death, because she would be writing from memory. 'Although she had not entered the house for a quarter of a century, every stick of furniture and every picture was accurately recorded in Vita's prodigious memory. Few objects had been shifted since her self-exile from Knole and I do not recall questioning a single reference to the contents in her text.'² The endowment for the house at the time was estimated, quite unrealistically, at £5000 a year: Lord Sackville offered to pay £3000 if the Trust could find the extra £2000. This was possible thanks to a proviso in Mrs Ronald Greville's will which permitted the Trust to make use of any surplus revenue from Polesden Lacey.

In 1947 the third Lord Leconfield handed over Petworth in Sussex with a £300,000 endowment and the Capability Brown park. He had been persuaded to take the decision by his nephew and heir, John Wyndham, later first Lord Egremont of a new creation as well as sixth Lord Leconfield – 'one of the bravest things that I've ever done. He could have struck me out of his will.'³ John Wyndham realized that after paying death duties he would never be able to preserve intact what he described as 'surely one of the most beautiful white elephants in Europe, an immense white elephant adorned by one of the most splendid collections of pictures to be found outside a public gallery.'⁴ Apart from its paintings by Van Dyck, Claude, Titian and other Old Masters, the collection is famed for twenty paintings by Turner, some of which had been commissioned by the Egremont of the day. The house itself, enlarged on a palatial scale in the late seventeenth century, is embellished by some of Grinling Gibbons' most elaborate carvings.

When his uncle died in 1952, Wyndham offered most of the collection to the government provided that it remained at Petworth. There ensued four years' haggling over which pictures should be included and at what valuation. After three years Crawford wrote to the Treasury:

> The N.G. [National Gallery] advises against accepting a number of pictures. . . . The suggestion that they should not be included would absolutely wreck the house, leaving a series of spaces on the walls. The suggestion is quite lunatic, quite indefensible, quite exasperating. . . . Owners won't face these obstacles and the years . . . of uncertainty

and misery. . . . If owners refuse to play – and if Wyndham
sells – no one can blame them or the National Trust.[5]

Fortunately Wyndham's patience lasted for another few
months and agreement was finally reached on the basis of a far
lower valuation than Wyndham had been advised to accept.
The Duke of Devonshire was less patient: he withdrew his
offer of Chatsworth to the government when arguments about
the value of the contents had gone on for two years.

The Trust was able to accept several important houses, for
which little or no endowment was forthcoming, by leasing
them to local authorities or to the government. This was the
solution reached in 1947 for Lyme Park in Cheshire, which was
given to the Trust by the third Lord Newton and leased to the
Borough of Stockport. Lees-Milne first visited the house – one
of England's greatest, he thought – in 1943. The ground was
covered in snow, the house intensely cold and Lord Newton
hopeless. 'The world is too much for him and no wonder. He
does not know what he can do, ought to do or wants to do. He
just throws up his hands in despair. The only thing he is sure
about is that his descendants will never live at Lyme after an
unbroken residence of 600 years.'[6] Lord Newton nearly abandoned his attempt to give away his property when two years
later yet another report was demanded. 'I am not prepared to
permit a further invasion of valuers, architects and surveyors
into the house. If they are not satisfied with Smith's report I
shall drop the whole thing. It is quite intolerable that an offer
of this nature should be treated with such suspicion and distrust. I feel much inclined now to sell out, when the time
comes, to the highest bidder.'[7] It must be doubtful whether a
bidder would have materialized for the vast house, Italianized
by the Venetian architect Leoni, who superimposed on the
Elizabethan core a great classical portico and an arcaded court-
yard more suited to Mediterranean sunshine than to the mists
of the approaches to Manchester.

The negotiations for Ham House and Osterley Park, west of London, were even more protracted. Ham House, on the banks of the Thames at Petersham, has escaped the decorative tastes of later generations and still retains much of its original panelling, velvet hangings, parquet floors and fine furniture. Built in 1610 and remodelled extensively for the Duchess of Lauderdale in the 1670s, her descendants, the Tollemache family, were still living there in the 1940s, even if it was a frugal and austere existence. Lees-Milne's description of the house during the war is one of his most poignant:

The grounds are indescribably overgrown and unkempt. I passed long ranges of semi-derelict greenhouses. The garden is pitted with bomb craters around the house, from which a few windows have been blown out and the busts from the niches torn away. I walked round the house, which appeared thoroughly deserted, searching for an entrance. The garden and front doors looked as though they had not been used for decades. So I returned to the back door and pulled a bell. Several seconds later a feeble rusty tinkling echoed from distant subterranean regions. While waiting I recalled the grand ball given for Nefertiti Bethell which I attended in this house some ten years or more ago. The door was roughly jerked open, the bottom grating against the stone floor. The noise was accompanied by heavy breathing from within. An elderly man of sixty stood before me. He had red hair and a red face, carrot and port wine. He wore a tail coat and a starched shirt front which had come adrift from the waistcoat. 'The old alcoholic family butler,' I said to myself. He was not affable at first. Without asking my name or business he said, 'Follow me.' Slowly he led me down a dark passage. His legs must be webbed for he moved in painful jerks. An ancient voice cried, 'Come in.' The seedy butler then said to me, 'Father is expecting you,' and left me. I realized that he was the bachelor son of Sir Lyonel Tollemache, the fourth baronet, aged eighty-nine. As I entered, the ancient voice said, 'You can leave us alone, boy!' For a moment I did not understand that Sir Lyonel was addressing his already departed son.[8]

Lees-Milne wrote ceaselessly to the local authorities, who dithered and procrastinated until Sir Leigh Ashton, director of the Victoria and Albert Museum, suggested that the government should buy the property and the museum should be put in charge. However, Sir Lyonel Tollemache objected to government ownership. Lees-Milne stepped in with a compromise: the house and gardens would be given to the Trust, which would lease them to the Ministry of Works. The Government would then buy the contents, for which the V & A would be responsible. This proved the basis for an agreement with the donor, but the Trust was not happy when Ashton proposed moving selected items to his museum. Crawford wrote to the Treasury protesting: 'The basis of our purpose and the purpose of the owners of these properties is that the contents should remain *in situ*.'[9] He gained his point.

Not far away, even lengthier negotiations were taking place between the Trust, the government and the ninth Earl of Jersey, the owner of Osterley Park. Lord Jersey had first approached the Trust in 1938, but he withdrew his offer after the house and grounds suffered damage during the Blitz and some of the pictures were lost at sea on their way to the island from which his title came. Bombs had fallen in the park, windows had been blown out of the house and the orangery had burnt down. In 1944 Osterley was again offered to the Trust and eventually, after the failure of negotiations with the Middlesex County Council, the offer came to fruition on the basis of arrangements similar to those agreed for Ham. Jersey wrote to Crawford, 'It is a great relief to me to feel that so much beauty is now in safe hands. . . . I was afraid that the place and its contents might be dispersed, something which I feel is almost sacrilegious. May I say how good Jim Lees-Milne has been – and patient too.'[10] Its value lay not only in the interiors and furniture designed by Robert Adam but also in the park, lakes and farmland which still give the impression of a country-house estate despite being cut off on the north by a motorway and on the other sides by suburban housing.

Soon after the war private gifts enabled the homes of two Prime Ministers to come to the Trust – Hughenden Manor in Buckinghamshire and Chartwell in Kent. Benjamin Disraeli, Earl of Beaconsfield, might have approved of the Trust's owner-ship. He had bought Hughenden with about 170 acres in 1848 and lived there until his death in 1881. Lord Ronald Gower recorded this account of his visit in 1880:

> Lord Beaconsfield talked in anything but a conservative sense as to the intolerable injustice of trying to keep people out of one's parks, especially when so near London as Hughenden is. He showed a great and good feeling about wishing to give as much enjoyment as possible to the hard and over worked classes as is compatible with private rights and seems to have a great contempt for the narrow, selfish views of many of the Tory and Whig landed proprietors, who make their class odious to the people by keeping them as much as possible out of their great demesnes and vast parks. 'I for one,' he said, 'cannot and will not do anything so absurd.'[11]

On Disraeli's death Hughenden was left to his nephew and in due course was vested in a charitable trust which passed it on to the Trust with the help of funds from the Disraeli Society. The study is still much as he left it and the house, furnished in dark heavy Gothic style, has many of his books and mementoes.

Hughenden feels like a museum compared with Chartwell, with its light airy rooms, the morning papers on the table and a half-finished canvas in the garden studio. Chartwell, Winston Churchill's home since he had bought it in 1922, was presented to the Trust by 'a number of fairly wealthy men [who] wish to benefit Mr X [Churchill] and preserve the residence as a national monument.'[12] The group (organized by Lord Camrose) bought Chartwell for £50,000 and handed it to the Trust with an endowment of £10,000, later raised to £35,000 on the understanding that Churchill would continue to live there for his lifetime and that thereafter the house would be kept as a memorial. Chartwell practically adjoins the woods of Mariner's Hill to the west and Toy's Hill to the east, which Octavia Hill had helped to preserve.

The houses accepted by the Trust during the five years after the war were only a proportion of those offered. From the beginning of the Country-House Scheme in 1937 until 1951, at least eighty houses were refused or withdrawn for financial reasons. Some of these were left empty and others sold for institutional use. Less than half of the 350 most important houses listed in 1939 were still lived in by the family ten years later. In the words of the 1951–2 annual report: 'Buildings on which our greatest architects, sculptors and painters have lavished their genius, and which stand high among our country's achievement, are today literally falling down; their irreplaceable contents, brought together by successive generations, are being dispersed; their gardens are overgrown; and the surrounding parklands of which they form the central and essential feature are becoming derelict.'[13]

Sir Stafford Cripps, Dalton's successor as Chancellor, was so concerned about the situation that in December 1948 he appointed a committee, chaired by Sir Ernest Gowers, a former chairman of the Board of Inland Revenue, 'to consider and report what general arrangements might be made by the government for the preservation, maintenance and use of houses of outstanding historic or architectural interest which might otherwise not be preserved, including, where desirable, the preservation of a house and its contents as a unity'. The Committee interpreted their terms of reference broadly to include the gardens and all the buildings and land forming part of the setting of the house.

The Trust was naturally one of the organizations invited to give evidence. Its statement was put together after much discussion and correspondence on the basis of a memorandum prepared by Lees-Milne and rewritten by Harold Nicolson (an elegant couple of authors for a statement of evidence). The Trust did not resurrect the 1936 proposal that owners should be helped to retain their own houses by means of tax concessions, assuming that this would be unacceptable to the government. But the Trust did reiterate the case for retaining the historic link between the houses and the owners in order to prevent what had been family homes from becoming 'mere lifeless museums'. The solution proposed was that the Treasury should continue to accept suitable inhabited houses with their estates in part payment of death duties and should offer them to the Trust on the understanding that the previous owners and their descendants might continue to live in them in return for paying rent.

The Executive Committee could not agree, and so made no recommendation, on whether houses no longer lived in by the family and not suitable for letting should be administered by the Trust or by the Ministry of Works. Esher contended that such 'museum' houses should go to the state. But Norman foresaw complications. 'How are we going to make the division when a house falls in? Consider a Montacute which we long hoped could be let as a house but which has turned into a museum. Cotehele, what is it, house or institution?'[14] Lees-Milne supported Norman. 'I am not in agreement with your suggestion', he wrote to Esher, 'that we compromise with the government over country houses. I do not think it at all easy to distinguish between houses inhabited by institutions or families. We have several occupied by both. They are doomed and there is no pretending otherwise. Do we surrender the houses when the families go?'[15]

There was also disagreement on whether the Trust should accept grants from the government, Nicolson putting the case against in a lengthy letter to Crawford:

I feel that the moment we accept the principle of being subsidised by the government, we lose our voluntary character and enter the corridor which leads to a government department. Once the Civil Service find that a so-called voluntary association is accepting subsidies, that association would result in a government department. I think the Committee should realize that if we commit ourselves on paper to [such] a suggestion . . . we lose our independence or what the Civil Service call our identity. . . . This issue must be faced, considered and discussed. There will be some who feel as I do that it would be better for the Trust to carry on on its own resources at the risk of losing new historic houses which we cannot afford to accept. There are others who will feel that the main purpose is to preserve historic houses, and that if the Trust cannot afford to do this they must say so frankly and allow the Treasury and the Office of Works to take on their functions.[16]

Crawford had some sympathy with Nicolson, writing a few months later, 'To ask the state for £100,000 would be to ask for nationalization, and to break our faith with every donor.'[17] But financial imperatives prevailed and the Trust recommended that it should be given an annual grant for the maintenance of properties transferred by the Treasury or inadequately endowed, and for the purchase of historic houses and estates. No sum was mentioned.

The Gowers Report was published in June 1950, eighteen months after it had been commissioned. The report paid tribute to the Trust's pioneer work and pointed out that it was the only means by which positive protection could be given to historic houses, whether inhabited or not. Starting with a masterly analysis of the problems of preserving country houses in a world where they could never again play their traditional role, the report proposed a balanced package of recommendations which, had they been adopted in full, would have enabled many more houses to survive intact with their collections and would have avoided the recurring crises over the break-up of individual houses.

The Committee adopted a twin-track approach. On the one hand they aimed to prevent houses from being abandoned by their owners, for 'the owner of the house is almost always the best person to preserve it'. They therefore proposed that the owners of 'designated' houses should receive relief from income tax for expenditure on repairs and maintenance; and relief from death duties on the house, contents and amenity land. But the Committee recognized that even with a more favourable tax regime many owners would not continue living in large country mansions and that it was necessary for the government to take some action. They recommended that Historic Building Councils should be set up as administrative agencies responsible to the Treasury with power to compile lists of 'designated' houses and also of their essential contents; to make grants or loans for the repair or upkeep of such houses; and, in case of need, to purchase and hold houses until a suitable use could be found.

On the day the report was published Crawford welcomed its proposals in a letter to *The Times*. But he expressed disappointment that it had not declared unequivocally that inhabited houses acquired by the nation should be vested in the Trust. Nor had it supported his view that chattels associated with historic houses should be accepted in payment of death duties, giving the surprising reason that few owners would sell, though it was known that the contents of Knole, Petworth and Cotehele were on loan to the Trust and would almost certainly have to be realized to meet death duties. The Trust's case for exempting chattels found favour with the Labour government, whereas the Gowers Committee's recommendation for tax exemptions to help what were perceived as a privileged class of owner did not.

Shortly before losing office in the 1951 general election the government published a draft Monuments and Buildings (Protection) Bill. This proposed giving grants to owners, though only a small proportion would have benefited compared with the number who would have benefited from tax relief. The Labour government, like its Conservative successor, preferred to tax and then repay a small proportion of the proceeds in grants. In addition it proposed the compulsory purchase of listed buildings if 'in danger of damage or destruction through neglect or injudicious treatment'; the listing of historic house collections (a step never taken); and more effective controls over the demolition and alteration of listed buildings. The Trust was pleased with the offer of grant aid, but told the Ministry of Works in private that owners would object to many of the other proposals, Crawford adding in a letter to the permanent secretary, 'There is much in them with which I personally strongly disagree.' The election of the Conservative government with Churchill as Prime Minister brought the Bill and these exchanges to an end.

The period of the Attlee government was the only time in the Trust's history when there was a real possibility of the state taking on the sort of direct responsibility for preserving both historic buildings and fine landscape that it exercises on the continent and in North America. This would have been in tune with the current political mood. Ministers were in the process of extending public ownership to the social services and the basic industries and after its wartime efforts the Civil Service was confident of its ability to take on almost any job. Senior civil servants believed that the Trust would not be able unaided to maintain indefinitely the historic houses it had already acquired, as they told Esher and Lees-Milne when discussing the possibility of grants. At that time the Treasury was more concerned with status than with the strictest control over public expenditure: the V & A's management of Ham House, Osterley Park and Apsley House in London had been sanctioned without difficulty and, when consulted about the future of Harewood House, the Princess Royal's home in York-shire, the Trust was advised that it would be more appropri-ately owned by the government.

Sir Edward Bridges put the official view of alternative approaches to the preservation of historic houses when he told the Trust about the Gowers Committee:

> It seems inescapable that the appointment of this Committee will raise issues of policy for the National Trust which are both difficult and important. On the one hand, if the Com-mittee were to recommend that the task of looking after those houses should be among the functions of the Trust, there would be a very difficult, perhaps impossible, financial prob-lem for you unless the government made the necessary funds available for preservation and upkeep of these houses. But the receipt of a substantial government grant might well prove to be inconsistent with the continuance of a satisfac-tory flow of voluntary contributions, and with continuing freedom from government control.
>
> On the other hand, if they were to recommend that the preservation and upkeep of historic buildings which may in future pass from private to public hands should rest not with the Trust but with some other body, either a government department or a new body in some degree of association with a government department, then one of the things which the Trust has set out to do up to the present would in future be done by some other body.[18]

Had Gowers' proposal for establishing Historic Buildings Councils as administrative agencies been adopted, Bridges' second alternative might have come about. Alternatively, had the Labour government's draft Bill been enacted, it might well have led to more public ownership as in France.

In the National Parks also the Trust might have been left with a lesser role had Dalton's intention to buy extensive areas through the Land Fund been carried out. In that case the Parks might have been owned by the government rather than being little more than enhanced planning authorities.

While so much attention was concentrated on country houses the Trust and the Royal Horticultural Society had launched a scheme to try and rescue some of the country's great gardens, which were falling into decay as a result of wartime neglect, lack of money and the difficulty of recruiting gardeners. The proposal came from the second Lord Aberconway, who had been president of the RHS since 1931 and owned a magnificent garden, Bodnant, in North Wales. Aberconway had inherited the garden through his mother's family and a fortune to support it from his father, whom he succeeded as chairman of John Brown & Co., the Scottish shipbuilding and engineering firm. These strands came together to make him both a prominent horticulturalist, interested in botanical experiments, and an autocratic industrialist with wide business connections.

On 25 November 1947 at a meeting convened at the Trust's office, Aberconway proposed that a joint committee should be formed to raise money and administer a few of the best gardens. The Trust of course already owned a number of outstanding gardens attached to country houses like Blickling, Stourhead and Cliveden. The new scheme was to take in gardens in their own right and these were defined as gardens 'of great beauty, of outstanding design or historic interest' or 'having collections of plants or trees of value to the nation either botanically, horticulturally or scientifically'. The Committee would also advise on the Trust's other gardens. Expert guidance had hitherto been somewhat lacking as Lord Rosse, vice-chairman of the Historic Buildings Committee, admitted in a lecture he gave to the RHS the following year. The Gardens Committee continued to provide overall supervision when Graham Thomas was appointed gardens adviser in 1954.

The Gardens Committee first met on 23 March 1948. Lord Aberconway took the chair and his fellow representatives from the RHS were Sir Edward Salisbury, director of the Royal Botanical Gardens at Kew, and Dr H. V. Taylor, horticultural adviser to the Ministry of Agriculture. From the Trust were Vita Sackville-West, the creator of Sissinghurst; Lord Rosse, a collector of exotic trees and plants for his garden at Birr Castle in Ireland; and Sir David Bowes-Lyon, the Queen's brother, a banker and also the owner of an outstanding garden.

An appeal was launched for the Trust's Garden Fund, but far more important in the long run was the support of the National Gardens Scheme. This had been successful in raising money by opening private gardens on occasional weekends in aid of the Queen's Institute for District Nurses. As the Institute was being wound up and incorporated in the new National Health Service, Vita Sackville-West suggested to its chairman that a proportion of the income from this source might go to the Trust's Garden Fund. The chairman of the Institute did not immediately warm to the idea but she was persuaded and in its first year's operation the expanded scheme collected £12,000, most of the proceeds going to the nurses. The scheme flourished and, while continuing to benefit various nursing charities, has become the principal benefactor of the Trust's gardens, contributing no less than £176,000 in 1989.

The first garden to come to the Trust in its own right was not one which had been abandoned during the war but Hidcote in Gloucestershire, cared for by its owner and never without its complement of five gardeners. Nor was it an historic English garden. Hidcote was barely forty years old and had been created by Lawrence Johnston, who was born in Paris of American parents and came to England as a young man. Johnston was one of the great artist–plantsmen of this century. Out of the bare fields of his Cotswold farm he made a wholly original garden, its vistas lined with high hedges which separate a succession of 'rooms', each with its own colour scheme and selection of plants. When his memory failed and he found the garden a struggle to look after, he decided to present it to the Trust in order to ensure its preservation. As an endowment he offered his house and a farm of 284 acres, let at £400 a year. This, plus whatever the Trust might receive in rent from a fully furnished manor house, was not going to pay for five gardeners. While the Executive dithered over the endowment Lees-Milne wrote to Bevir: 'I do think that the garden is of such exceptional importance we really must press the Committee again to say they will accept the devise in anticipation that the Gardens Committee will raise a fund. Major Johnston will not live long.'[19] The Executive Committee was persuaded and Hidcote was accepted. The following year Bevir received a rather different memorandum from Martineau: 'Please do not think me pernickety – I observe in the Executive Minute of 16th July that Hidcote would be accepted subject to satisfactory financial arrangements. I can find nothing on the file to show that satisfactory arrangements have been made. The Garden Fund will have to grow with remarkable rapidity.'[20] Martineau's fears were not unfounded, for within two years expenditure at Hidcote outweighed income by £1650.

The second garden to come to the Trust in its own right was Bodnant. Aberconway was anxious that, having devoted so much time and work to the garden, it should go on after his death. He offered an ample endowment, subject to the provision that his family would have the right to act as managers. This was not unreasonable since the garden had been started by his grandfather, had been developed by his mother and himself, and was to be further enlarged by his son, who in his turn became a long-serving president of the RHS and chairman of John Brown & Co. The four generations of Aberconways were almost matched by three generations of head gardeners in the Puddle family. Bodnant descends from a series of broad formal terraces looking towards Snowdonia to the stream where exotic trees, rhododendrons and other shrubs flourish in the deep ravine.

The Gardens scheme was launched thirteen years after

Lothian's call for the Trust to come to the rescue of historic country houses. Its aim was less ambitious – to save at least a few irreplaceable gardens at a time when year by year they were disappearing. By 1990 the Trust owned 160 gardens of all types and sizes, most attached to historic houses. They included nearly thirty National Plant Collections and had become a magnet for visitors, many of the Trust's most popular properties in 1990 being gardens.

Gardens, however, formed only a fraction of the 41,000 acres acquired by the Trust between 1946 and 1950. About a quarter of the total was accounted for by the estates which came with historic houses, and some 7000 acres were given or bequeathed in the Lake District. A rare gift in southern England was the Slindon estate, still the Trust's largest holding in what Sir Arthur Hobhouse had hoped would become the South Downs National Park, but which has never been designated as such. Slindon's towering beech trees, pointing straight to the sky, were among the finest in the country until they were felled by the great storm of 1987. Above them the land rises to a Roman road cutting across the downs and to Glatting Beacon, from where there are breathtaking views over the Weald and the Channel. The estate, including its mansion and part of the village, was left to the Trust by Mr Wootton Isaacson, who in 1944 announced that he wished to hand it over before he died. With the promise of £55,000 for repairs and an endowment, the Trust quickly accepted the offer. But Mr Isaacson was persuaded to cut this sum by more than half, leaving an inadequate endowment and no money for repairs. After his death in 1948 there ensued an unfortunate and semi-public row between his sister, who was living in the Dower House, and the Trust on the question of the rent she was charged to meet the cost of repairs. Feeling that the Trust had been unfairly accused, Hubert Smith, the Trust's chief agent, was prompted unsqueamishly to comment, 'I hope the Trust will never weaken because they anticipate discontented relatives and other vultures who seem to hover around rich people when they are dying.'[21]

During these years when the Trust's head-office committees and senior officers were preoccupied with the consequences of the Labour government's legislation and decisions on new acquisitions, the recently appointed land agents were struggling with the management of the Trust's agricultural estate. The full-time staff numbered only about thirty in 1950 and they were assisted by ninety-five local committees. The war had left an enormous backlog of maintenance on buildings, walls, fences, paths, gates and stiles and in the woodlands, some of which had been clear-felled. Many properties had been scarred by army occupation, which had left behind the usual detritus of concrete emplacements, empty huts and barbed wire. The 1947–8 report recorded, 'The work of reinstatement goes on slowly, but not nearly as fast as we would wish. Labour, materials and money are all scarce to say nothing of such matters as permits.'[22] In the Lake District, Acland quickly took up the grants offered under the 1945 Hill Farming Act for comprehensive farm improvement schemes. The 1949–50 report records a substantial increase in expenditure on repairs and improvements to farms, and expresses the hope that 'funds will allow this programme . . . to be continued not only in the interest of agriculture, but also because, if properly controlled, a prosperous agricultural community can do much to preserve the beauty of the countryside'. Standards were high: 'In carrying out these improvements the Trust has always so to design and site them, at increased cost, as not to injure the beauty of the land and buildings it is preserving for the nation.'[23]

Costs in fact proved to be much greater than had been anticipated, largely because of the rise in wages. Farms on estates like Blickling, Killerton and Stourhead had become liabilities rather than sources of income, requiring large injections of money to provide the cottages with electricity and running water. It was clear that few historic houses had been taken on with realistic endowments.

In the country houses minor repairs had been postponed and occupation by troops had inevitably left some damage – at Blickling the RAF had broken window casements, smashed old crown glass and forced doorlocks. Although Cliveden, Wightwick and some of the Trust's recently acquired houses had been well maintained, others were in need of major structural works and their ornamental garden temples were often in a state of dereliction. Nonetheless the public expected the houses to be opened. Even before the end of the Japanese War in the Far East, letters to *The Times* were complaining that the Trust had not advertised where and when the public could visit its properties.

During the early months of 1946 Lees-Milne chuntered round the country in his rickety old car, engaging caretakers, rearranging furniture and preparing guidebooks. Inevitably there were crises: the day before Charlecote was due to open neither the guidebooks nor the British Legion guides had turned up (and when they did their 'information' was embarrassing to listen to). Despite the lack of staff, eleven historic buildings were opened during 1946 and nineteen more during 1947. And despite petrol rationing visitors began to flock in – 120,000 in 1948, 433,000 two years later.

In April 1948 Crawford went on one of his rare tours of inspection. On his return he sent some crisp notes to Esher on what he had seen and heard. Montacute was 'becoming a very good show place' while Barrington was 'all rather bogus'. At Stonehenge the 'barbed wire fences [are] very offensive. Make more effort to do away with as many as possible. They disfigure the greatest prehistoric monument of the world.' At Stourhead, Sir Henry Hoare's nephew, who was living in the house, told him that the reorganization of the contents was turning it into a museum, whereas it should remain as far as possible a house – an expression of the taste of the period.

Lord Crawford noted, 'An 1830 inventory of each room exists; it would be better to stick to this rather than place things where they might perhaps look better. We should also show the family as well as the house.' Several of the donors complained, as Crawford reported: 'There has been some ill feeling [at Stourhead]: not surprising when there was a proposal to take half of a huge set of Chippendale chairs to Lyme. I don't suppose we have been too tactful. Lees-Milne had told me that Hoare was interfering with the management. This may have been so but it is the wrong point of view. We must not forget the donor – we tend to do so.' At Lacock Abbey he found both house and grounds shabby, and Miss Talbot had told him that the Trust was not managing the Estate's affairs too well; Knollys and Buxton are 'very polite, very nice but of course quite incompetent. Any farmer could get the better of them. It would be prudent to seek her [Miss Talbot's] advice more than we do.' West Wycombe was no better. 'What squalor, we should be ashamed of ourselves. The grounds are very ill kept and the outside of the house is in very bad condition. I came in for the full blast of complaint.'[24]

The effort to catch up with repairs exposed the inadequacy of the Trust's funds to meet the obligations of maintaining both its farms and its historic houses. So long as De La Warr was chairman of the Estates Committee (1950–1 and 1955–70) the agricultural estate had priority: cowsheds and milking parlours, piped water and electricity, and the replanting of woodlands. 'The year's financial deficit is formidable,' states the 1950 report. The Jubilee Appeal was almost exhausted and, although income from visitors and from members was increasing, the total revenue was quite inadequate except in years when there was an unexpectedly large legacy. At the end of 1950 the reserves stood at £122,000, the total income was £145,000 and the deficit amounted to £56,000. This was after spending £250,000 on repairs and improvements in the previous two years and before embarking on the dry rot and leaking roofs of some of the newly acquired country houses. Not surprisingly the Trust waited with anxiety to hear whether financial assistance would be forthcoming from the new government.

CHAPTER EIGHT

Western Approaches

IT WAS not long before the new Conservative government took action on the Gowers Report. Although R. A. Butler, Chancellor of the Exchequer, proved no more willing than his Labour predecessor had been to grant tax reliefs for privately owned historic houses, the Trust was well pleased with the Historic Buildings and Ancient Monuments Act, which received royal assent in October 1953.

This set up separate Historic Buildings Councils for England, Scotland and Wales, as advisory bodies, and established a system of financial aid which lasted for just over thirty years in England and to this day in Wales. In England the Minister of Works was empowered, after consulting the Council, to make grants for the upkeep of buildings of outstanding historic or architectural interest, together with the contents and adjoining land, and to buy such buildings or to assist local authorities and the Trust with their purchase. After Harold Nicolson had been to see Butler, the Trust was successful in extracting a concession that the Minister might also make capital endowments for houses given to the Trust. This, however, proved to be an empty victory, for the Trust was offered no endowments in England or Wales until the National Heritage Memorial Fund was set up nearly thirty years later. The Labour government's proposal to require consent for the demolition or alteration of listed buildings was dropped, not to be enacted until 1968 when, after the demolition of a number of important buildings, the idea was revived by a later Labour government.

The Historic Buildings Council for England soon established a position of sufficient influence to ensure that its
recommendations were rarely overturned. This was largely
due to its chairman, Sir Alan Lascelles, formerly private secretary to successive monarchs (mainly King George VI), a man of
military appearance, authoritative manner and strong opinions.
It was also due to its stable and distinguished membership,
which included a leading architect, an art and architectural
historian and MPs from both sides of the House of Commons.
No fewer than three of its members also served on Trust
committees – the Earl of Euston (later the eleventh Duke of
Grafton), Christopher Hussey and W. M. F. Vane, MP. This
connection helped to establish a close relationship which
lasted until the HBC was incorporated into English Heritage in
1985.

In Wales, the seventh Marquess of Anglesey ensured that
there were equally close links through his membership of the
Historic Buildings Council from its inception until 1992 (chairman from 1977) combined with his membership of the Trust's
Regional Committee from 1960 (vice-chairman 1975–85). In
1976 he gave his own house, Plas Newydd, where the sun-
drenched Italianate mural by Rex Whistler looks across the
Menai Straits to the rain-drenched panorama of Snowdonia,
and where the first Marquess's wooden leg lies in its glass-
fronted case.

In 1951 Robin Fedden succeeded Lees-Milne as historic-
buildings secretary of the Trust and it was his responsibility to
submit the first applications for grants. More of an organization man than his predecessor, he was that unusual creature, a
competent aesthete, and he became a dominant influence
during the twenty-three years he held the post. Born in 1908,
he was brought up largely by his grandparents until the age of
ten while his father, a landscape painter, and his mother, an
American writer, were abroad. In 1920 they settled in France,
where Robin joined them for holidays from Clifton College and

Cambridge. In 1937 he became cultural attaché at the British Embassy in Athens and in 1940 went to live in Egypt, where he taught English at Cairo University. He travelled widely in the Middle East and, a conscientious objector, he served for a time with the Friends Ambulance Unit. Returning to England after the war he was appointed curator of Polesden Lacey and then regional representative in southern England. Lees-Milne summed up his quality as historic-buildings secretary: 'He was essentially an eclectic connoisseur in the eighteenth-century tradition. This was his strength. Above all he knew which experts to approach for opinions. In a trice he picked their brains and immediately acted on his own assessment of their (often conflicting) advice.'[1] But he was not interested in nineteenth- or twentieth-century buildings, refusing to contemplate, for example, the last of the linen magnates' mansions on the outskirts of Belfast. His complex personality combined a steel-like quality with the sensitivity to soothe the most dissatisfied of donors and to ease the problems of staff, but he never had the committees in his pocket as did Lees-Milne. An enthusiastic, almost a fanatical, mountaineer, he wrote several travel books, poetry and a novel. His finely honed style and his gift for evoking landscape could have made him a full-time writer. He set high standards for the Trust's guidebooks and other publications and wrote its first history, *The Continuing Purpose*, published in 1968.

Fedden lost no time in applying for help from the HBC. In the first two years the Trust received nineteen grants, ranging from a few hundred pounds for conserving pictures at Charlecote and Lacock to thousands for repairs at Knole, where the programme continued until the 1980s.

Large grants made it possible for the Trust to accept several houses which were in urgent need of repair and could not otherwise have been afforded. Among them was Ashdown, high on the Berkshire downs, a Flemish-looking house with a steep roof built about 1660 by the first Lord Craven for the

Winter Queen. Unfortunately the contents associated with her had already been sold, but some of the portraits have since been returned and subsequent purchases of surrounding land have safeguarded the views from the roof, which is reached by an immense staircase and crowned by a golden ball. Fedden informed Esher of the precautions he was taking and also of the local gossip:

> In order to avoid another Nether Lyppiat with huge architect's costs and no house at the end of it I have got from Lady Craven's architect the drawing for such work which she had obtained when she decided to abandon Ashdown to its fate. Needs £10,000 spent on it. It appears that Lady Craven, an American, married when sixteen in 1894, and that her father had secured for her absolutely Ashdown and 30,000 acres. Her local reputation is that of a thoroughly bad landlord who is not in the least interested in public opinion. She sells constantly, never visits and dislikes her grandson, the present Lord Craven.[2]

A very different landlord was Ralph Verney, whose house, Claydon in Buckinghamshire, was also on the verge of dereliction. It had been left empty after 1928, apart from a period housing children from evacuated schools, and presented a fearsome task for Verney when he returned from war service. He and his wife had just about made the house habitable and had held the first of what became regular summer concerts when they could no longer 'ignore the fact that the lead on our roof was becoming more and more a sieve: tin and china baths were less and less able to contain the raindrops: you could smell the mushrooms even through tobacco smoke in the tower, and one day looking up at the dome of the Grand Stairs we saw mushrooms growing down as well as up, while little worm casts littered the floor in the Chinese Room'.[3] They called in the HBC, who awarded £40,000, the largest grant for repairs made during its first decade, and this enabled the Trust to accept Verney's offer of the house and four acres of garden.

The HBC also played a crucial part in the complicated negotiations for transferring five houses accepted by the Treasury and offered to the Trust without adequate endowments. Grants were made to meet the deficit on upkeep but these came as annual allocations, which have gradually fallen in value, rather than as capital sums. Pre-eminent among the five houses was Hardwick Hall in Derbyshire, which was built by Bess of Hardwick in the late sixteenth century and still retains most of her tapestries and pictures, some hanging in the Long Gallery, which has been described as the greatest room in Europe. The present Duke of Devonshire was burdened with especially heavy death duties when, from the point of view of family settlements, the inconveniently timed death of his father followed that of his elder brother in the Normandy campaign. The idea of offering that 'more glass than wall' phenomenon to the Trust came to him while he was waiting for a train on Chesterfield station just over the hill. His grandmother, Duchess Evelyn, continued to live there, writing to Lord Rosse, 'I love this place and enjoy showing it to people who appreciate the things in it. Also it is a comfort at this rather sad moment when it will cease to be a family home to know that it will be in safe hands.' She added, 'I am thankful that the National Trust is probably taking Hardwick over. It seems much the best thing that could happen to it. What I do very greatly regret is that no member of the family is likely to live there in the future, not because of the Trust but because Andrew [Devonshire] could not manage an extra place and there is no one else anyhow at present.'[4]

The *Field* evidently did not share the Dowager Duchess's opinion of the people who came to visit, commenting in a leader, 'This newspaper doubts whether the people as a whole are yet qualified to benefit from such enlargements of the human spirit. . . . It is for this once proud but now coarsening nation that Hardwick shall exist.'[5]

Under Esher's forceful leadership the Trust acquired nineteen country houses during the years 1950–5 and made many more agreements in principle which fell through, mostly because adequate endowments were not available. Esher's determination to press ahead with such empire-building caused some worry to Crawford and he occasionally upset others by his railroading methods. Martineau objected to the terms agreed with some owners and periodically threatened to resign on the issue.[6] One unsatisfactory arrangement concerned Nostell Priory in Yorkshire, where the Trust was obliged to pay for an indoor staff of eleven and also for the insurance and conservation of the magnificent Chippendale furniture still owned by the family. But these concessions were due to the Private Bill required to break the entail being amended by his fellow peers, who thought that Lord St Oswald needed more protection. It was not until 1985, when the collection was bought for the Trust by the National Heritage Memorial Fund and the house was leased back for the family to manage, that a satisfactory arrangement was made. It was a remarkable feat for the Trust's small staff to take on so many large properties in such a short space of time, a feat made possible only by the fact that in most cases donors and their agents continued to run them very much as in the past.

Nostell was the last major country house to be acquired while Esher was chairman of the Historic Buildings Committee. He was succeeded in 1955 by Michael Parsons, sixth Earl of Rosse. Born in 1906 Rosse, like most of the Trust's committee chairmen at the time, was a substantial landowner, educated at Eton and Oxford. He was not, however, a conventional country gentleman. His principal home was Birr Castle, County Offaly, in Ireland. He was a man of high intelligence and exceptional taste. For twenty-one years he guided the Committee firmly but persuasively 'with a sure knowledge of the Trust's aims and purposes and a clear understanding of its integrity'. Frequent tours gave him a detailed knowledge of the properties and his diplomacy was appreciated not only by the staff and committees but by the donors, whose problems he was adept in sorting out. Coming from a family with a long tradition of public service and scientific interests, 'he maintained his ancestors' distinctive absence from the Irish sporting field; he neither rode, nor fished, nor shot, but his contribution to various branches of culture and the preservation of old and beautiful things and the creation of new ones was not slight'.[7] He presided over institutions on both sides of the Irish Sea, holding office as Pro-Chancellor of Trinity College, Dublin, president of the National Collections of Ireland and president of the Royal Horticultural Society of Ireland as well as being chairman of the Standing Commission on Museums and Galleries in London.

His interests were shared and encouraged by his wife Anne, the daughter of Colonel Leonard Messel, who bequeathed his garden, Nymans in Sussex, to the Trust. She entertained lavishly at Birr, at Womersley in Yorkshire and at their London house, but she was much more than a sparkling society beauty. Like her husband, she inherited a passion for gardens and they spent their honeymoon collecting exotic plants for Birr and Nymans. Michael Rosse became a noted dendrologist, his interest aroused by the gift of some rare trees on his twenty-first birthday. But design he left to Anne, who had some of the artistic talent of her brother, Oliver Messel, the theatrical designer. She later managed Nymans for the Trust.

The Rosses also shared a commitment to historic preservation. Both were early members of the Georgian Group and Michael Rosse served as its chairman for twenty-two years. Founded in 1937 under the aegis of the SPAB, its three talented but disputatious leaders soon found William Morris's doctrine of repairs too much of a straitjacket and split off to form a separate organization. The style enjoyed by the Georgian

Group's aesthetes and bright young things with a taste for champagne and caviar would in any case not have been compatible with that of the more earnest members of the SPAB accustomed to beer and sandwiches. Although Georgian buildings had attracted academic attention since the turn of the century, they were still dismissed by the public in general as dull square boxes. The Group's objective was to change this attitude and to halt the long list of demolitions exemplified by the loss of Devonshire House, Regent Street and the Adelphi.

Twenty years later Anne Rosse took a leading part in founding the Victorian Society. The first meeting took place in the house in Stafford Terrace, Kensington, which had been taken in 1874 by her grandfather, the illustrator Linley Sambourne, and which still retained the interiors he had created. She and John Betjeman were elected vice-chairmen and Esher, then aged seventy-six, presided over the new organization. The perception of Victorian architecture went through a cycle of disregard similar to the Georgian one, but at an accelerated pace. Admiration for the confident civic palaces and industrial temples of the nineteenth century had given way to the view that any building described as 'Victorian' was thereby damned, until public appreciation began to revive just as Norman Shaw's Scotland Yard and Gilbert Scott's Foreign Office were threatened with demolition.

By the late 1950s the immediate country-house crisis was passing. Some houses had been saved from the demolition contractor by the Town and Country Planning Act and had been converted to new uses. Others had passed to the Trust, most with their contents on loan even if not given or transferred in lieu of death duties. Above all, there was a new feeling of hope among the younger generation of owners. They were encouraged by government help with repairs (grants through the HBCs for England, Scotland and Wales totalled more than £1 million by 1957) and their personal finances were improving as agriculture prospered, the yield on investments increased and tax rates on high incomes showed a reduction, even if modest. The mood of greater confidence gradually strengthened during the years of Conservative government. Many owners were prepared to adapt themselves and their houses to a simpler style of living and were anxious to make an income from visitors.

There was nothing new in opening large country houses to visitors. In the eighteenth century the housekeeper at Chatsworth had instructions to show people round, and by 1849, the first summer after the opening of the railway from Derby, there were 80,000 visitors. What was new was to charge a significant fee for entry and to introduce shops, restaurants and other attractions in order to help pay for the house. The Duke of Bedford was one of the first to realize that 'Most people enjoy these places considerably more if their basic needs are catered for and good loos and food are available, and if the place is clean.'[8] He provided these and other attractions for marketing his 'stately home'. Woburn had been offered to the Trust and accepted subject to finance in 1955, the same year in which the Duke had opened it on his own initiative. After being shown Petworth and Uppark by Lord Euston, then the Trust's part-time representative, who hoped to persuade him of the advantages of being relieved of financial troubles while continuing to live in the house, he became openly critical of the Trust's 'many fine, sad houses'. When Crawford visited Woburn four years later he reported:

> The Duke dislikes quite passionately the idea of transfer. What he has seen of other NT houses strengthens his dislike. The arrangement by which the family lives on sufferance in a house which they have given away is artificial and undignified; the occupant is a mummy preserved in someone else's museum. He is just enjoying running Woburn the way he wants to. The Trust would be compelled if we accept Woburn to put an end to much that is happening there. What is happening the public enjoys and comes to enjoy in ever

increasing numbers. To stop this would incur much public odium, no less damaging than the press attacks.[9]

Crawford wisely called off the negotiations.

Most owners did not wish to emulate Lord Bath's pride of lions at Longleat, nor were they able to develop a major motor museum as did Lord Montagu of Beaulieu. Some continued to find life difficult. Lord Brooke, the owner of Warwick Castle (later sold to Madame Tussaud's), complained, 'It is not really funny when those who are struggling with the problems of maintaining much of our great architecture are caricatured as coroneted and berobed clowns pocketing ten-penny pieces.'[10] But Evelyn Waugh thought that he had been unduly pessimistic in the picture he presented in *Brideshead Revisited*. 'It was impossible to foresee in the spring of 1944', he wrote some years later in a prologue to a revised edition,

> the present cult of the English Country House. It seemed then that the ancestral seats which were our chief national artistic achievement were doomed to decay and spoliation like the monasteries in the sixteenth century. So I piled it on rather, with passionate sincerity. Brideshead today would be open to trippers, its treasures rearranged by expert hands and the fabric better maintained than it was by Lord Marchmain. And the English aristocracy has maintained its identity to a degree that then seemed impossible.[11]

Most of the fourteen historic houses which passed to the Trust during the years 1956–60 were dependent on substantial financial help from the government. An exception was Waddesdon Manor in Buckinghamshire, with its superlative collection, one of the most munificent bequests to a public body since the Wallace Collection in 1897.

Waddesdon was built between 1874 and 1889 for Baron Ferdinand de Rothschild by the French architect, Destailleur. Crawford and Esher did not like this enormous pastiche of a

French château. Esher also hated French furniture, but Crawford was persuaded by the quality of the collection: 'I don't see how anyone who has seen it can fail to be quite overwhelmed by the beauty of the objects and the arrangement unless of Marxist principles he disapproved of luxury, splendour and glitter.'[12] The objects included furniture made for the French royal palaces, the finest Sèvres china and French, English and Dutch paintings and drawings, as well as a magnificent library, rare manuscripts and collections of glass, majolica and other works of art.

At the end of the war when the evacuees had departed, Mr and Mrs James de Rothschild, who had no children,

> were confronted with the problem of what to do with the house. Shortages of every sort in the post-war period included a shortage of personnel and the idea of reinstating Waddesdon for us two old people seemed out of the question. But something had to be done: we could not leave all the contents of the house in a few rooms, stacked up to the ceiling. . . . It was then that the solution of the National Trust began to germinate in my husband's mind and I cannot say how thankful I am that it did.[13]

The Executive Committee accepted the house on condition that it was accompanied by the principal contents of the show rooms and 'a very liberal endowment' (appropriate as James de Rothschild had been a Liberal MP until 1945). The Treasury at first objected that the £600,000 proposed was more than would be needed for upkeep and should therefore be subject to tax. Eventually they gave way but the transfer had not been completed by the time James de Rothschild died, leaving to the Trust the house, all the contents of the ground floor and 120 acres, provided that it was declared inalienable within three months.

Crawford decided that special arrangements would be needed to manage Waddesdon: a board of trustees was therefore set up with Mrs de Rothschild as chairman and five others, two nominated by the family and three by the Trust. The trustees have maintained the house to 'the Waddesdon standard' (well above what the Trust could afford), acting with a unique degree of independence, since the 'liberal endowment' proved inadequate to meet the cost of annual upkeep let alone major repairs, without the injection of large subsidies from Rothschild family charities. Mrs de Rothschild found that 'the sudden change . . . from an existence which had been totally free from taking decisions of importance was traumatic' and leant heavily on Crawford's advice. In February 1959 she apologized: 'I have a horrid feeling I am going to write you the longest [letter] ever and indeed a prospect from which even the hardest eye might recoil'[14] – there followed eight long pages.

Mrs de Rothschild continued to employ the same experts as before to advise on the collection. In other houses responsibility fell to the area representatives (by 1960 there were ten) who relied on help from the national museums, the Courtauld Institute and other bodies. This led to some complex arrangements. For example at Arlington Court in Devon, where the collection was more curious than important, the Natural History Museum advised on the shells and minerals, the Maritime Museum on the two hundred model ships, the Society of Pewter Collectors on the pewter, and the Victoria and Albert Museum on the textiles. One of the Trust's serious weaknesses was the lack of catalogues for paintings, other works of art and libraries. On one occasion in 1955 Crawford wrote to say, 'I've heard from several sources some disturbing things about our way of dealing with archives and books. I mean to go into this with you – it is something in which I am interested and know something.' But there was no money for looking after the archives and Fedden defended his practice of depositing the larger collections of archives with the County Record Offices, which had been set up after the Second World War. In 1959 the Pilgrim Trust gave a grant towards cataloguing the Trust's libraries, which made it possible to make a start with preparing card indexes for the earlier books at Blickling, Charlecote and Attingham.

A year later Fedden reported on the arrangements at Ham and Osterley, which were costing the Victoria and Albert Museum £25,000 a year for indoor staff and showing, without including the garden or the repairs:

> At Ham there were twelve warders on duty summer and winter, although during the winter months they only have some 2000 visitors. The number of warders cannot be reduced as owing to trade union regulations one warder may not cover more than a specified number of rooms and seasonal employment is impossible. The ordinary cleaners cannot dust the tops of the picture frames and anything high on the walls as they are not allowed to do work above their reach when standing on the ground. Another trade union regulation. Two women are unnecessarily engaged at Ham in shifts to sell tickets because no warder is allowed to handle money. Again trade union regulations. No flowers were allowed to be supplied from the gardens but only from Hyde Park, which was naturally expensive. The cleaners are not allowed to touch any 'work of art', e.g. legs of an Adam table at Osterley. Consequently a craftsman has to be sent to clean such a table from the museum. A craftsman cannot go without a foreman. This usually means a motor car in addition.[15]

Fedden was convinced that the Trust should not enter into such an agreement again, and that it should reconsider its policy of not taking on 'museum' houses. Apart from other advantages the Trust could have managed with three or four people for each house, working under a single curator.

Fedden's responsibilities extended beyond England to Wales and Northern Ireland. Eight years earlier Clough Williams-Ellis had commented, 'Wales is miserably under-represented by Trust possessions' because in Wales there is 'the mistaken notion that just because a place may be remote and wild so it will always be safe ... a still surviving mistrustfulness of English institutions run from London and generally anything not arranged for Welshmen by Welshmen for themselves after long discussion – in Welsh! ... few notable buildings of the middling manor house sort and fewer still comparable to the great country houses.'[16]

An Advisory Committee for Wales had been set up in 1945 with Lord Merthyr as chairman and Williams-Ellis among its members, and a full-time area agent had been appointed, but in 1948 there were only 500 members (3 per cent of the Trust's total).

Wales ceased to be 'miserably represented' when in 1951 the vast neo-Norman castle of Penrhyn, outside Bangor, together with nearly 40,000 acres of the Snowdonia National Park was transferred to the Trust on the initiative of Lady Janet Douglas-Pennant via the Land Fund procedures. The Trust was doubtful whether it could manage what is still its largest estate and also whether the castle was of sufficient architectural merit. The Duke of Wellington thought not, but the Historic Buildings Committee was convinced by Lees-Milne's report recommending acceptance of the castle, which had been built between 1830 and 1847 by Thomas Hopper, as 'in relation to its setting ... a highly successful piece of architecture. It absolutely fulfils what it was designed to do, namely to present a bold impression of romantic intensity.'[17] The case was strengthened by Lady Janet Douglas-Pennant's willingness to lend the interesting 'Norman' furniture and the best private collection of paintings in North Wales.

The estate includes ten of the main mountain peaks of Snowdonia and also Lake Ogwen and the upper Conwy Valley. But the farms had been neglected in favour of the shooting, most of the tenants were elderly and if the property was ever to reach modern standards it would take years of work. John Tetley, newly appointed area agent for North

●

Wales, was given the task. 'In 1951 we took over fifty-five inhabited holdings varying from a few acres to 500 and averaging about 100 acres. Of these fifty-five farms not one had a water supply that was working satisfactorily. Each had its individual spring or stream of soft acid water which had corroded the piping. Only five of the farm houses had a hot and cold water supply; two had WCs, one of which was outside.' The farm buildings, access roads, cottages and woodlands were in equally poor order.

Ten years later Tetley described what he had accomplished with the help of the estate foreman, who was Welsh-speaking, and with funding from government grants: 'Today all except two farms have had their water supply improved or renewed; forty-four have proper hot and cold water and twenty-six have bath and WC as well. Although domestic improvements do not produce any income to help a tenant pay his rent the fact is that today a good farm with a bad house is more difficult to let at a poorer rent than a poorer farm with a good house. Wives will not accept the old discomforts.' The cowsheds, sheep dips and fencing had also been brought up to the required standard. 'Woodlands are slowly being mended. . . . We are trying to produce very mixed woodlands, partly in order to retain some of the old hardwood colours, but also because it is felt that the market for small parcels of soft woods fifty years hence is far from certain.'[18]

The key to Tetley's achievements was his success in winning the confidence of the tenants and the estate staff, who at first resented being managed by a charity based in London. Tetley had the imagination to understand their feelings. He felt 'pretty sure that once it was realized that our aim was not only to be conscientious landlords but to respect, protect and enjoy the truly Welsh ways of the estate, as part of the Trust's responsibility in holding it, the initial suspicion of us as an enemy to be defeated might be changed to a realization that we were an ally'.[19]

179

While the negotiations over Penrhyn were proceeding, two important houses were offered to the Trust in South Wales. Negotiations for the acquisition of Tredegar Park in Monmouthshire broke down because the trustees were unable to provide an endowment and did not wish it to be maintained by the Ministry of Works (it was later acquired by Newport Borough Council). Picton Castle in Pembrokeshire was accepted by the Executive Committee, but again the offer fell through (the house is now owned by a family trust). Other houses such as Hafod, Llangibby Castle, Glanusk Park and Llanwern Park were demolished; the Trust owns no large houses in South Wales and few survive with their contents.

A year after Penrhyn, in 1952, the great sandstone stronghold of Powis in mid-Wales was given to the Trust. The fourth Earl of Powis, then aged eighty-one, had written to say that he was considering handing over the castle, with the park, 351 acres, the furniture in the public rooms and a £75,000 endowment, on the understanding that his successors could live in the rest of the house. The Trust had no hesitation in accepting the house, with its fine contents amassed over 500 years of continuous occupation and its renowned terraced gardens cascading down to the valley below.

Negotiations became deadlocked over the need to obtain the approval of Lord Powis' heir and the difficulty of securing an adequate endowment. But, as a result of the good offices of Lady Berwick of Attingham, agreement was reached in October on the basis that the Trust would have the house and gardens and a £100,000 endowment, but not the contents, which would remain on loan; the family would continue to live in the castle. At the end of the month Lord Powis suffered a collapse but his signature was vital because unless death duties could be saved the family would not have been able to provide the necessary endowment. Rathbone wrote to the agent on 28 October, 'I realize that Lord Powis is now a very sick man and I am so anxious that there should be no possible suggestion from any quarter that the Trust is presuming on his age or ill-health to get this transfer through quickly. Could you give me some assurance that no question of this can arise?'[20] A cousin, the then Bishop of Norwich, was given the power of attorney, and, without the matter going before the Finance Committee for approval, the property passed to the Trust on 6 November. Lord Powis died three days later.

The transfer of Penrhyn and Powis alerted Welsh opinion to the National Trust's activities in the Principality and there were some voices raised in favour of an independent Trust for Wales. In January 1955 a group of Welsh MPs considered asking the Home Secretary to investigate the possibility.

The Advisory Committee for Wales was not a particularly representative or active body. Tetley recorded, 'It was composed of eminent personages in the Principality, few of whom could speak Welsh. It met twice a year; once in the south and once in the north, and received a report from the agent, who tended to report only what he wanted them to hear and show them only what he wanted them to see.'[21] Occupied as Chairman of Committees in the House of Lords and as chairman of the Pembrokeshire quarter sessions and other organizations, Merthyr had little time to spare for the Trust.

The Executive Committee realized that something had to be done and done quickly to meet the desire for a more effective Welsh presence. The Advisory Committee was re-named the Committee for Wales and two MPs were invited to join – Roderick Bowen from the Liberals and Goronwy Roberts from Labour (though Kenneth Robinson, then the Labour MP on the Trust's Executive Committee, was asked if this was really necessary, and when Roberts dropped out a few months later he was not replaced). A more active chairman who could also serve on the Executive Committee was sought and Goronwy Rees, the brilliant but mercurial Welsh-speaking Principal of Aberystwyth University, seemed an ideal choice. But after little more than a year he resigned when his period at

Aberystwyth came to an untimely end. He was succeeded by the more respectable figure of Colonel J. C. Wynne-Finch. In 1968 when the Welsh membership stood at 2500 Fedden's verdict was that 'there existed in Wales neither the public interest nor the financial backing to support an independent Trust'.[22]

Twenty years earlier Matheson had taken a similar view of the prospects for a separate Trust in Northern Ireland, believing that, 'it would certainly be small and would probably be boycotted by Catholics'. During the war, when the Northern Ireland Committee had asked for additional powers to take decisions on several matters including the acquisition of new properties, he had gone to Belfast to discuss matters. Having previously told Zetland that the Committee 'has been divided against itself and does little', he was unexpectedly impressed by its new young and active members and by the recent increase in Trust membership from under twenty to more than eighty. He refused to delegate authority over new acquisitions, but he agreed that the Committee should have power to sign short-term tenancy agreements and promised to consider whether the Committee might introduce by-laws and oppose parliamentary Bills at Stormont without waiting for approval from London. Permission was not, however, granted.

In 1948 the situation was transformed by the appointment of an exceptionally effective chairman, Randal McDonnell, thirteenth Earl of Antrim. He had joined the Committee in 1938, serving in the navy during the war, and he held the office for seventeen years before becoming chairman of the whole Trust in 1965. He lacked the aesthetic taste or knowledge of Crawford, but he had a good eye for a painting and, with John Betjeman as a friend, was one of the first leaders of the Trust to be interested in Victorian architecture. A large untidy man, his cheeks always wreathed in smiles, he dedicated his life to the Trust. His warm, down-to-earth qualities were exactly what was needed to cope with the problems that erupted in the

1960s. Born in 1911, he was, like Rosse, educated at Eton and Oxford and divided his public work between Ireland and England, where he was chairman of a group of London hospitals. He married a Catholic and was for that reason unable to play a full part in the life of Ulster, just as Rosse, a Protestant, was also a little detached from the public life of the Republic.

Two years after Antrim became chairman Ulster acquired its own separate Land Fund. This was modelled on Dalton's Land Fund for Great Britain, but it had one critical difference – it could be used to provide endowments. Antrim immediately wrote to Rosse, 'The Ulster Land Fund is exciting in that there is £1 million for anything the Minister of Finance approves and I think that he will agree to finance Castle Coole, Mount Stewart Gardens etc. All we have to do is to put up cases to him and make them convincing enough to satisfy him that the case is good.'[23]

Endowments were certainly needed. The financial position of many owners was even more difficult than in England, the Irish Land Acts having resulted in large estates being broken up which left the owners, unless they had other resources, with little more than the park and the home farm to pay for the upkeep of the mansion. Most of the large houses offered to the Trust in the 1950s came with barely enough land to pay for the salary of a gardener, let alone for a new roof. Antrim's optimism was not misplaced. By March 1956 the Ulster Land Fund had dispensed £260,000 with a further £60,000 promised, about half of which had been spent on providing endowments for properties which the Trust would otherwise have had to refuse.

The 1950s were a *decennium mirabile* in Northern Ireland. Nine outstanding properties came to the Trust, all with financial help from the Ulster Land Fund. Castle Coole in County Fermanagh, a superb classical house built in the 1790s by James Wyatt for the first Earl of Belmore, was the first. The seventh Earl of Belmore offered the house to the Trust on condition that he could live in one wing and retain ownership of the contents, which he would leave on loan. The Northern Ireland government released £50,000 to cover the cost of purchase and repairs and to include £35,000 for an endowment. This 'generous and far-sighted gesture' was heralded by the *Estates Gazette* as the first occasion on which a capital sum had been placed at the disposal of the Trust and exactly the sort of initiative recommended by the Gowers Committee. It was not long before other owners were persuaded by Antrim, who could talk to them as a familiar neighbour, that the future of their properties could be ensured only by transfer to the Trust. In 1953 Castle Ward, a bizarre house, part Palladian, part 'Strawberry Hill' Gothick, on the shore of Strangford Lough, was transferred with an endowment from the Land Fund. In 1954 Florence Court, the early-eighteenth-century home of the Earls of Enniskillen, was given by the son of the fifth Earl and endowed by the Land Fund. In 1955 the garden of Mount Stewart and in 1956 the garden of Rowallane also came through the Land Fund.

Relations with the donors who continued to live in their houses were not without difficulty. It was Antrim who had to appease the elderly Lady Bangor when, living on her own at Castle Ward, she wrote to complain:

> Could you come to my rescue and try to get Commander White [secretary to the Northern Ireland Committee] to see to the water supply for this house. All he does is to get an architect and tell me the main water supply will be here in no time . . . whereas I have on reliable information it cannot be done for well over a year . . . and then he says, Lady Bangor you must wait, and that does not seem very kind in my old age. Do you know that I have not had a bath in my house for months, I have to motor to my neighbours for one. Then the sanitary arrangement????? I know I could get the sanitary people to demand this.

Finally, as an apologetic aside she adds, 'Please forgive this lengthy letter but I have been living in great discomfort for some months now.' In another letter she writes, 'I hope you will understand what a worry it is to see everything falling into decay all round. I am very worried about the stable yard as I do not think the chimneys will stand another winter and are about to fall and if they fell the garages will be badly damaged.'[24]

Mount Stewart was another property where there were difficulties with the donor's family. The gardens, with their rare trees and imaginative lay-out, were transferred to the Trust through the Ulster Land Fund by the widow of the seventh Marquess of Londonderry and her daughter, Lady Mairi Bury. The memorandum of wishes was based on that for Bodnant, though the circumstances were quite different in that Lord Aberconway had given his garden with a large endowment and both he and his son were highly respected in the horticultural world. Antrim had refused to accept the house at Mount Stewart on the ground that 'we should stick to our standards – the highest'.

Within a few years there was a disagreement over Lady Mairi Bury's part in the management of the property and the employment of gardeners. Antrim kept Rathbone in touch with events:

> I tried to explain to Lady Mairi Bury that the National Trust would be taking an unacceptable risk if they shared their gardeners with her or allowed their gardeners to be controlled by a head gardener who was not a National Trust servant. I also tried to point out that the fact that she ran a market garden was one of the reasons that made that type of compromise impossible. . . . I feel convinced that the principle for which I am standing over this matter is of profound importance. It is very easy to criticize the National Trust by saying that we are using public funds to maintain old and decaying families in their homes when they can no longer afford to do so themselves. As you know better than I do that

is the Beaverbrook line and it can sound plausible in England. If we over here are using grants from the Ulster Land
Fund and are not particularly careful in the control of our
money we could not only be justifiably criticized ourselves
but we would also let down the Ministry of Finance which
has been so immensely generous to us during the last ten
years.[25]

Rathbone pointed out in reply that 'She [Lady Mairi Bury] was
one of the donors of these gardens. I remember well that it was
the late Lord Aberconway who played a big part in persuading
Lady Londonderry to give us these gardens and I have no
doubt that she thought and Mairi Bury thought that they
would have a say in the management of them. Indeed we are
on this committed by the memo of wishes.'[26]

Eventually a compromise was reached whereby the Trust
controlled the expenditure and the gardeners would not be
shared but Lady Mairi Bury could co-operate with the Committee in the management. Relations improved and in 1976 the
house, where Lord Castlereagh of the Vienna Congress had
grown up, was given to the Trust with an endowment provided by Lady Mairi Bury and the Ulster Land Fund, her father
having earlier told Lees-Milne that he did not care whether the
whole place went to the Trust or was left to fall down.

It was hardly surprising that owners had grounds for complaint because until 1960 Antrim and his Committee had to
manage the scattered estates with the help only of a part-time
secretary. The Committee had been strengthened by the
appointment of a woman Independent Member of the Stormont Parliament, the editor of the *Belfast Telegraph*, the curator
of the Armagh County Museum and the young partner of a
leading firm of Belfast solicitors. This was Charles Brett, who
brought his energy, initiative and intimate knowledge of
Ulster's buildings to the service of the Trust in Northern
Ireland for thirty-seven years. Insisting that if more members

were to be recruited there must be more publicity, he wrote a *Guide to Properties* himself, commissioning the illustrations from James McIntyre, a talented young local artist, and organized a series of lectures by John Betjeman among others. Later his acerbic wit and delight in stirring things up enlivened the proceedings of the Council. Rosse, Fedden and others came over to Ulster to advise on difficult decisions: whether the house of Lord Brookeborough, the Prime Minister of Northern Ireland, should be accepted, even though rather mediocre, for fear of causing offence (it was refused and no offence was taken); how Florence Court should be redecorated after a devastating fire, and so on.

By 1964 when Antrim retired from the Northern Ireland Committee to take over the General Purposes Committee in London the Trust owned a wide variety of properties. They included Mussenden Temple, a ruinous Graeco-Roman rotunda on a cliff overlooking the Atlantic; the picturesque village of Cushendun, with cottages designed by Clough Williams-Ellis; and the Giant's Causeway, the famous basalt rock formations, together with a ten-mile cliff path along the North Antrim coast. On his retirement Antrim reported to Rathbone: 'Clanwilliam has taken my place; he is first rate in spite of the fact that he is another Earl!'[27]

CHAPTER NINE

Gardens and Quirkeries

IN ENGLAND the Trust once again turned its attention to campaigning against the vulgarization of the countryside and building on fine stretches of coast. With the spring *Bulletin* of 1957 were circulated particulars of *Counter Attack*, 'a series of practical proposals for dealing in sane and appropriate fashion with the development of the English landscape in town, village and open country'. Even the *Estates Gazette* expressed alarm at the failure of the planning legislation to control the effects of the economic revival and the housing programme: 'If what is called development is allowed to multiply at the present rate,' stated a leader in the same year, 'then by the end of the century Great Britain will consist of isolated oases of preserved monuments in a desert of wire, concrete roads, cosy plots and bungalows.' It added, 'there will be no distinction between town and country; the end of Southampton will look like the beginning of Carlisle. The parts in between will look like the beginning of Southampton or the end of Carlisle.'[1]

Most of the land which came to the Trust in the 1950s – 122,000 acres, more than in any other ten years except 1975–85 – was not situated between Southampton and Carlisle. Much the largest area was in Snowdonia, the 40,000-acre endowment for Penrhyn Castle. But there were other houses with parks and gardens which were nearer the centres of urban population – Attingham in Shropshire, the Vyne in Hampshire, Hardwick in Derbyshire and Tatton in Cheshire. So, too, was one of the largest devises ever received by the Trust – the Buscot and Coleshill estates in Oxfordshire and the

Bradenham estate in Buckinghamshire, amounting in total to 8000 acres. These were left by Ernest Cook who used the fortune derived from the sale of the family travel agency business to buy first Montacute and the Bath Assembly Rooms, which he had already given to the Trust, and then seventeen traditional rural estates. In 1945 Matheson was told that they would all be left to the Trust, but after he retired relations became strained. Hubert Smith and Lees-Milne had to deal with Captain John Burrow Hill, the somewhat overbearing agent to whom the shy and reclusive Cook entrusted his affairs, and who was thought by Cook's family to treat the estate as if it were his own. The break came because Cook wanted the whole of his estate declared inalienable, whereas the Trust felt that only about half was of the requisite quality. In 1951 Cook cancelled his proposed gifts except for those which he was covenanted to assign and announced his intention to found a separate charitable trust bearing his own name.

A number of extensive estates, and some smaller ones, were transferred to the Trust through the Land Fund during the 1950s, including 5600 acres of the Marsden Moor Estate near Huddersfield, 6500 acres of the Derwent Estate in the Peak District and 7000 acres in the Lake District – five separate lots in 1959–60 alone. There was also a continuing stream of gifts and purchases of land from the proceeds of bequests and donations, particularly in the Lake District where Acland was active in attracting support and in Cornwall, where Michael Trinick took over as area agent in 1954.

A spectacular coastal acquisition was Brownsea Island. Almost land-locked within Poole Harbour it has all the magic of an island far more remote, and perhaps for this reason was used by Baden-Powell for his first scout camp. Until 1927 it had been a well-kept estate of arable fields, water meadows and woodlands. Then it was bought by Mrs Mary Bonham-Christie, who was intent on banishing human contact and allowing nature to take its course. Before long much of the island had become a jungle of rhododendrons, inimical to other plants and to wildlife, and had been devastated by fire. But it remained a refuge for red squirrels and a sanctuary for wildfowl. In 1960 John Bonham-Christie was left the island by his eccentric and solitary grandmother, together with a bill for £200,000 in death duties. He put the island on the market, aware that developers would think it a perfect place for holiday camps and expensive homes.

But he had reckoned without Rathbone, who had noticed the announcement in *The Times* and immediately wrote suggesting that the island should be offered to the Trust via the Land Fund. Despite the support of the local planners and the Nature Conservancy, the deal would have foundered had not the John Lewis Partnership, still run in the paternalistic style of their idealistic founder, agreed to take a lease on the mostly nineteenth-century castle which lay on the eastern tip of the island as a holiday centre for their staff. The Trust needed to raise £100,000 for upkeep and repairs but Rathbone uncharacteristically urged that the island be accepted forthwith on the grounds that it would be an opportunity for positive preservation and 'would do us a great deal of good'.[2]

At this juncture Helen Brotherton of the Dorset Naturalists' Trust, who lived on the mainland opposite the island, offered to help with the appeal and then to persuade her members to look after the nature reserve. The island was an instant success, attracting 50,000 visitors in its first year. Helen Brotherton was one of several keen naturalists who were attracted to the Trust by its work for nature conservation. She joined the Executive Committee and the Council and later became a highly successful chairman of the Wessex Regional Committee.

Towards the end of the 1950s the Trust embarked on an entirely new departure – the restoration of the Stratford-upon-Avon Canal. This was due to John Smith, who had acquired an interest in old buildings as a schoolboy and had been appointed to the Historic Buildings Committee in 1952 at the age of only twenty-nine. He joined the Executive Committee and the Council in 1961, becoming chairman of the General Purposes Committee when Esher retired. A brilliantly original and highly idiosyncratic man, he became through his Manifold and Landmark Trusts a great benefactor of the Trust.

Smith was fascinated by early industrial structures, especially the canals, which had been quietly declining since the advent of the railways. By the mid-twentieth century, hidden away from roads in the countryside and behind back streets in the towns, they were in danger of being abandoned. Spending an August Bank Holiday weekend exploring canals and meeting only one canoe, he realized that this secret world could become an amazing leisure resource. In a persuasive letter to Crawford in March 1958, he explained why the Trust should interest itself in canals:

Tow-paths, which are not rights of way and are therefore lost on abandonment, provide long, beautiful uninterrupted footpaths.

Fishing. The amount of coarse fishing which goes on in canals would be unbelievable were we not ourselves getting over £2000 a year for the coarse fishing at Clumber.

Scenery etc. Canals bring to parts of the country which do not otherwise have it, waterside scenery – gay boats, water plants, water birds and light-reflecting water itself. Furthermore, canals stretch green fingers right into the centre of the very worst towns: in many towns the canal is the only thing of beauty.

Boating. More and more people have boats.

Safety. Canals not only take people off the road; but every barge carries as much as between seven and twenty five-ton lorries.

Silence. The fact that canals are most numerous where the population is densest increases all these advantages.

Historic interest. The canal system is the largest

eighteenth-century remains in the country. Furthermore, there is no system like it – nothing so old or so unaltered in any other country in the world.

Architecture. Most canals were made when, for the last time, beauty and utility were unselfconsciously combined. Consequently, one can travel by water through mile upon mile of harmonious environment – the navigation works, the lock cottages, the wharfs and warehouses, and, most beautiful of all, the aqueducts and innumerable bridges – all in harmony with each other, and with the countryside.[3]

Crawford was convinced by Smith's eloquent arguments and the next Executive Committee agreed that the Trust should go ahead subject to finance. Smith hoped that the initiative would divert the Trust's attention from dukes' seats towards unfashionable initiatives of benefit to ordinary people, and would attract volunteers to carry out the hard labour, then a new idea for this type of project. By 1958, when an application for closure was being considered, the Stratford-upon-Avon canal had become a *cause célèbre*. Protesters packed into the town hall to urge the Trust to intervene, but nothing could be done without a Private Act of Parliament. Once this was passed the Trust took the canal on a five-year lease with an option on the freehold if the canal was put in working order.

The canal passes along three aqueducts and under twenty-eight Georgian bridges on its fourteen-mile course from rural Warwickshire to the edge of Birmingham. Hundreds of thousands of tons of mud were removed, sixty-five lock gates replaced, undergrowth cleared and the bridges and tunnels repaired. Almost the greatest triumph was the response by volunteers to pleas for help. Hundreds turned up at weekends and for holidays to clear silt and fell trees; the Youth Hostels Association lent accommodation; the Territorial Army sent engineers; prisoners from Winson Green gaol in Birmingham took part; and contractors provided free advice and equipment. All were represented when the Queen Mother reopened the canal in June 1964.

The following year the Trust duly acquired the freehold in the hope that some time in the future ownership of the canal could be passed to a more suitable authority. But twenty years later no successor had been found and Smith's expectation that the canal would cost the Trust no money had proved wildly optimistic. A public appeal had raised £64,000 for the initial repairs whereas by 1982 running costs amounted to £40,000 a year and £1 million was needed for further repairs. The Trust proposed a scheme for the British Waterways Board to act as agent but no agreement was reached because the Trust refused to allow navigation rights on a stretch of the River Avon that connected with the canal and passed through the park at Charlecote, arguing that this would threaten the security of the house and the serenity of the park. Finally in 1987 an agreement was reached whereby the British Waterways Board accepted responsibility for managing and keeping open the canal on condition that the Trust provided £300,000 a year for five years as a contribution towards an estimated £3 million repair bill. It had been a successful even if expensive exercise and had done something to demonstrate that the Trust's concept of historic monuments embraced more than country houses.

Canals were not the only industrial monuments which Smith persuaded the Trust to consider during the early 1960s. At his suggestion the Executive Committee agreed to accept, subject to finance, the earliest surviving railway bridge in the world, abandoned railway lines, a Victorian gas works and the Abbeydale Steel Works in Sheffield. But even Smith was obliged to recognize that in these cases the costs would have been too high for a charity with so many widely scattered properties. All the Trust could do was to awake the interest of the local authority or some other body. One industrial monument the Trust did accept was the Conwy Suspension Bridge in North Wales, which had been built by Thomas Telford in 1826 and was threatened with demolition.

The rapid expansion in ownership made necessary some increase in staff and expenditure on property maintenance, and in the early 1950s this led to substantial deficits. In 1952, Rathbone insisted on a temporary standstill on all but the most essential work at properties dependent on General Funds (many properties were financed by their own 'Defined Purpose Funds'). Rathbone stood firm despite complaints from Knollys to Esher:

> I am writing after the last financial revelations and stormy meeting. Jack's idea is I know all-round economies and realization of everything possible such as sales of timber. I on the contrary feel that this may be quite a mistake. So far we have a good reputation as a landlord but it will be a great mistake for our goodwill if we become a stingy bankrupt organization which refuses to do the simplest things. There is a great danger that we shall throw away our goodwill, cut down trees and spoil places only to find we are in the same pickle again. We need to face a government grant of at least £100,000. I dread being known as a rotten old bankrupt charity instead of the best landowners in the country.[4]

Although the government grant did not materialize, the financial position eased after 1953 when the Council was allowed to vary the minimum subscription (immediately raised from ten shillings to one guinea) and to recover the tax on subscriptions paid by covenant. In addition, the Trust, in line with other charities, was given power to invest moneys as they saw fit instead of being limited to trustee stocks. This led to some increase in investment income and a measure of protection from inflation. Membership grew from 23,400 in 1951 to 92,000 in 1960, raising the income from subscription from £19,500 to £83,700. There was also an increase in the number of visitors to properties for which a charge was made: from 540,614 in 1951 to 727,000 in 1955 and more than a million in 1959, yielding an increase in income from £146,000 to £360,000. The only sources of income which showed no consistent upward trend during the 1950s were donations and bequests, which varied unpredictably from £50,000 to £160,000 a year.

Local authorities were increasingly drawn to help the Trust, perhaps because 'Where the preservation of historic buildings and the countryside is concerned, the conscience of the nation has been touched.'[5] The 1950–1 annual report expressed thanks to no less than seventy-one local authorities, most of which had made grants towards the upkeep of open spaces or local landmarks – for example, Hardcastle Crags near Bradford, and Kinver Edge, the sandstone ridge in Staffordshire, where twenty-one local authorities co-operated. Local authorities also came to the rescue of two country houses. Washington Old Hall, the thirteenth-century home of George Washington's ancestors, could not have been accepted unless the local district council had taken a lease, nor could Tatton Park with its fine collection and beautiful garden, if the Cheshire County Council had not done likewise.

Another source of income was timber sales from the 30,000 acres of woodland: in 1953, for example, £78,000 was raised, but £42,000 of this had to be set aside for replanting. A solution to the cost of replanting woodlands, many of which were still suffering from wartime felling and neglect, was to enter into an agreement with the Forestry Commission and so benefit from their grants. But the Trust was determined to ensure that it remained the sole arbiter of what was or was not considered detrimental to amenity and it was some years before a deal was agreed by which deeds of dedication were set up under the 1947 Forestry Act enabling the Trust to receive grants for Clumber, Blickling and other places.

These brief facts and figures give some idea of what a mammoth task the Trust had taken on in managing and

195

opening such a vast landed estate and such a diverse collection of historic sites. Many of the properties were run down and did not match the Trust's claim that 'It is always the National Trust's endeavour to maintain its holdings worthily,' nor the glowing descriptions given in the guide books. Visitors were sometimes disappointed, as was one who wrote to the *Daily Telegraph* in June 1957, pointing out that at Claremont all that was to be seen of the Grotto, marked on the map as a cascade and said to be associated with Princess Charlotte, was 'a jumble of smashed masonry, scattered boulders and rank weeds with a trickle of muddy water'.[6] Near by at Clandon Park in the upstairs rooms, 'the wallpapers are faded and stained, the carpets threadbare or ragged, the plaster cracked and the paintwork dirty'. Many years would be required to bring those houses and parks which had been starved of money for decades up to a reasonable standard.

The Beaverbrook press pursued a different line of attack: the Trust 'was formed originally to preserve open spaces for the working people of Britain. But it has moved away from that practical idealism. Now it is increasingly concerned with enabling privileged families to remain in subsidized occupation of them.'[7] This was part of a broader onslaught under the headline, 'Why should we let them cash in on history?', asking why the Duke of Marlborough was paid to restore the roof of Blenheim and the Marquess of Bath to eradicate beetle from Longleat. The most sustained attack on the Trust was reserved for its handling of Cliveden. It was claimed that it was not opened frequently enough and that Lord Astor, whose family had long figured on Lord Beaverbrook's blacklist, was benefiting at the taxpayer's expense. When the article was drawn to Antrim's attention he replied:

> I was very interested to read the cutting from the *Sunday Express* and I shall certainly keep it as the point of view of the article is one that we should always bear in mind. On the other hand there is really no need to take the Beaverbrook

press too seriously on this subject as all preservation and the National Trust in particular is a personal hate of Lord Beaverbrook himself, stemming from the occasion many years ago when a caretaker of a National Trust property refused to allow him to come in without paying his entrance fee – not a very distinguished motive.[8]

This story may or may not have been true, but in any case a do-gooding institution run in part by landed aristocrats was not likely to appeal to Lord Beaverbrook.

Far more serious than occasional pinpricks from the press was the failure to adapt the structure and staffing of the Trust to cope with its rapid growth. Between 1949 and 1954 over which period the Trust acquired 168 new properties, including forty-eight historic buildings and 84,000 acres, and trebled its membership, the staff was increased by only fourteen, from fifty-nine to seventy-three. Under-staffing was not the only problem: there was no clear chain of command; too many decisions were taken in London; working relations between area agents and representatives were often strained due partly to the loose definition of who was in charge of what; there was too often a breakdown in communication with donor families; and there were too many head-office committees. All this led to muddles and mistakes.

In 1954 in reply to a memorandum prepared by Rathbone, Hubert Smith, the chief agent, confirmed that all was not well:

I have no hesitation at all in saying that the job has got too big for Gibbs [the deputy chief agent] and myself to tackle efficiently. This is serious, because it is quite useless to attempt to solve problems that crop up at properties with which one is totally unacquainted. There is often no continuity in the progress of negotiations for a new property [at Ickworth six people were involved]. There is far too little consultation between the various departments at head office. Confusion often arises because too many people have a finger in the pie.[9]

Despite memoranda by Rathbone and a report from a small committee of Executive Committee members, little was done beyond abolishing three head office committees. The most significant change and one with far-reaching implications for the future was the appointment of two regional secretaries – Michael Trinick for Cornwall and Carew Wallace (who had worked in the Colonial Forestry Service and was active in the SPAB) for East Anglia and the northern home counties. They were intended to 'be countrymen of wide experience of the problems with which the Trust is concerned, not only of general administration, but particularly of taste and amenity'.[10] This attempt to find or breed hermaphrodite figures, training land agents to become aesthetes and vice versa, did not work. However, the two regional secretaries were the forerunners of the regional directors, who would later assume general management responsibilities. Despite the success of regional committees in Wales, Northern Ireland and the Lake District, an extension to other parts of the country was not followed up.

The long-serving chairmen of committees – Crawford, Esher and Rosse – were not interested in the dull question of administration nor did they show any awareness that a constitution designed for a few hundred members was quite unsuitable for almost 100,000. Members could vote for the Council only by attending the annual general meeting in person. Very few bothered to do so (about 350 in 1960). In practice the elected members of the Council were chosen by the Executive Committee. Apart from those members of the Council nominated by outside organizations, the Trust's Council and Executive Committee were a self-perpetuating oligarchy, as Antrim later admitted. But they were vulnerable to a takeover should a small group of dissatisfied members succeed in summoning their supporters to an annual general meeting.

The 1960s were perhaps the most creative and certainly the most controversial decade in the Trust's history. Enterprise Neptune, the campaign to protect the coastline, was launched and the structure of the Trust was completely overhauled, but only after a row which reverberated through the press for months (see Chapters 10 and 11). And the Trust was embroiled in disputes over two issues which have continued to erupt from time to time: threats of compulsory purchase to inalienable land; and the management of wildlife, in this case the culling of seals on the Farne Islands.

It was also the decade when the consumer society broke out from post-war austerity: washing machines and television sets filled the shops and their ownership spread from the middle classes to the better-paid manual workers. A more visible sign of increasing affluence was the jump in the number of cars, from two million in 1950 to over eight million in 1964. Macmillan's theme for the 1959 election, 'You've never had it so good', was reflected in the Consumers' Association, whose magazine *Which?* was founded in 1957 to guide shoppers through the bewildering maze of glossy new products; by 1965 the Association's membership was more than double that of the Trust. For young people it was the decade of motor scooters, the Beatles and the expansion of secondary and further education.

Crawford at Balcarres was not tempted by the new dishwashers (although he often did the washing up himself) and rarely witnessed the streams of cars penetrating the Lake District or North Wales or carrying their families to country houses and gardens. But John Smith, younger and working in London, who took over from Esher as chairman of the General Purposes Committee in January 1961, was more in touch with changing lifestyles. With Crawford in London only once a month and not easily available for consultation on the telephone, this was a key position. Once described as 'the only rich, intelligent eccentric left in England', Smith was seen as a great hope for the future, possibly even a future chairman. It was an imaginative move of Crawford's to try and anchor him to the Trust at the relatively young age of thirty-seven. After obtaining the approval of Esher and Nicolson, Crawford wrote to consult Rosse:

> I think JS is a person whom everyone likes, committees and staff, and that is important. I think he would do the difficult negotiations and deal with the personal problems of donors (which Oliver [Esher] has done brilliantly) very well indeed. I think that he would bring new ideas to us all and that we would enjoy the meetings. I think . . . I think . . . in fact I've thought a lot and long to hear what you think of the idea.

Rosse approved.[11]

From his new position of influence John Smith tried to reform the organization. Seven years earlier he had already outlined what he thought should be the Trust's objectives: 'The Trust exists to take the place, when it cannot be avoided, of enlightened private owners. If it is to do its job and keep its independence the Trust must be, and appear to the public to be, as unlike a government organization as possible.' He suggested that it should be based on three principles: '1. Each property should be dealt with by as few individuals as possible. 2. As many decisions as possible should be dealt with on the spot by the man on the spot. 3. There should be no haggling over trifles.'

To implement these principles he suggested that the functions of head office should be restricted and more responsibility delegated to area agents, but there should be no regional organization. On this occasion he ended his note modestly: 'Of course I have no thorough knowledge of the existing administrative organization, so what I say may be bilge – or too little of a change to solve the problem.'[12]

But it was more than a year after taking over before he was able to assemble his co-chairmen to discuss the question. 'It was like playing chess by postcard with a man in Papua.'[13] Part of the problem was the head-office committees. In a lengthy memorandum to Crawford, Smith pointed out that there were eight main committees with 101 members between them, not counting fifty seats on the Council. The multiplicity of committees consumed a great deal of staff time, caused long delays and wasted more money than was realized. The committees dealt not with separate functions, but with separate aspects of the same function: gardens could not be separated from taste, which was the province of the Historic Buildings Committee, and should not be separated from Estate Management; none could be separated from finance, public relations or policy. As a result the secretary, chief agent and historic-buildings secretary attended most meetings. Smith found the General Purposes Committee, which was intended to be the driving force of the Trust, lackadaisical. In 1963 it met on nine occasions at eleven o'clock in the morning. After an hour and a half members would drift out to lunch, leaving an unfinished agenda.

Some of the senior staff shared John Smith's dissatisfaction. Trinick wrote to Rathbone saying that, while the administration of the Trust had been improved by the recruitment of more professionally qualified staff, decentralized management, planned investment, a modern accounting system and proper rates of pay, the committees' efficiency had sunk to a low level. 'I think that nearly all of us feel that the Trust ought to be controlled by amateurs and that its ideals should be maintained and prosecuted in this way, but only if the amateur control is efficient and keeps in touch with men and events. This has not been the case for some time.' He went on to say that the proliferation of committees and the bickering between them were causing dissension among the staff: 'A substantial body of the agents, for example, look on the accountants as pernickety and unrealistic, on the representatives as vague and ineffective and on the publicity people as worthless. I am convinced that the fault lies in the committees who are similarly critical of and jealous of each other and some members of which, both wittingly or unwittingly, encourage the staff to follow suit.'[14]

Both Trinick and Smith were in favour of reducing the number of committees and having one small effective executive. But there agreement ceased. Trinick wanted fewer matters referred to committees and the staff left to get on with day-to-day affairs. Smith, on the contrary, wanted the General Purposes Committee to meet once a week and exercise direct control. In the short run Smith's view prevailed, and the General Purposes Committee agreed to meet once a fortnight, but in the longer run it was inevitable that the committees should limit themselves to policy and that its execution should be delegated to the staff.

The immediate difficulty was that Smith had lost confidence in Rathbone after several muddles had culminated in a complaint from Lord Aberconway. Writing to Crawford, De La Warr supported Smith in this dispute, adding that this was not the only incident in which the staff had obstructed the Committee's wishes. He went on to say, 'Jack is extremely good in his way and is completely dedicated to the Trust, but we are now so big that we cannot run it in the rather friendly family affair way that has been so successful in the past.'[15]

Relations between Smith and Rathbone went from bad to worse, Smith becoming impatient in his handling of Rathbone, who thought that Smith was gunning for him. The situation might have been sorted out, Smith believed, had Crawford not been so far away: 'It is the greatest possible shame that you are not nearer at hand. . . . You and I have amazingly similar views on the role of the Trust . . . and, with quite different characteristics, I think we should make a strong combination, but we never see each other and that is that.'[16] A week after receiving this letter Crawford wrote to consult Rosse, the colleague with whom he was most closely in touch:

I find Rathbone tiresome in many ways but I do not find him as completely and hopelessly inefficient as John S: do you? He misses opportunities, e.g. conferences called (apparently) by Prince Philip about which I knew nothing and all sorts of tiresomeness (I could curl up when he says 'Bless you: look after yourself, we need you so much!'). But I do think that John exaggerates and that he fails to see any of the good side, which is very important. They are both rather neurotic and both work themselves into a frenzy of nerves and overwork: but both have qualities of such value to the NT that I wish we could make the best of them – we are not at present.[17]

Norman recalls that Crawford came to see him one day and said, 'Rathbone, Fedden etc are going to quit unless J.S. goes. And I [Crawford] shall quit too.'[18] Crawford did not like harsh words or actions but he made sure that it was Smith not himself who went. Summing up what had happened, a colleague concluded that what Smith wanted to do for the Trust was very necessary, but 'he had no velvet glove on his iron and eccentric fist'.

In April 1964 Rathbone suffered a nervous breakdown and the following month Smith resigned as chairman of the General Purposes Committee, though he stayed on with the Trust, becoming deputy chairman in due course. Crawford hoped that, when Rathbone returned to work some months later, under Antrim as chairman of the General Purposes Committee, all would be well. Others were less optimistic. Field Marshal Sir Gerald Templer, who was a member of the Executive Committee, doubted whether anyone who had had a bad nervous breakdown would be fit to take on such a responsible position again. Chorley also thought that the job was too big for Rathbone and added, 'In my experience this type of patient often breaks down again.' They were to be proved right. Meanwhile only a few of Rathbone's closest colleagues and the leading members of the committees were aware of what had happened.

Despite these problems, the Trust continued to add to its properties, though not at the rate of the late 1940s and the

1950s. The increase in acreage amounted to 90,000 (about a third was in the Lake District) compared with 125,000 acres in the 1950s. The Lake District has always represented the heart of the Trust, both emotionally and in the size of its holding, but its properties took time to build up. In 1945, after fifty years, it owned only 3.5 per cent of the area (15,000 acres plus 12,000 under covenant), well below the Forestry Commission, the water boards and several private landowners. By 1990, however, it owned more than 25 per cent of the National Park (165,000 acres, plus 12,000 under covenant and a further 17,000 leased) and had become much the largest landowner. The two most extensive properties were high up on the fells – 16,800 acres of the Lonsdale Commons, including the Langdale Pikes leased in 1961 from the seventh Earl of Lonsdale for thirty-five years or until his death; and 30,500 acres of the Leconfield Commons around Wastwater and including the bed of the lake, transferred in 1976 by the first Lord Egremont through the Land Fund. This latter acquisition was important, not only because of its size (it was the second-largest property ever to come to the Trust), but also because it ensured that Wasdale remained one of the few unchanged dales in the area and it enabled the Trust to resist further water extraction from the lake.

Cubby Acland retired as regional agent in 1973, having done a great deal to enlarge the Trust's holdings. He also continued his programme of helping the tenant farmers to enjoy a reasonable standard of living: by the early 1960s most of the sixty hill farms had indoor plumbing, electricity, better systems of sheep-handling and improved farm buildings. In 1964 he set out his thought on what ought to be the size and shape of the Trust's farms in the Lakes:

> The Trust's policy is to build up its farms and its stocks of sheep, sometimes by amalgamation and sometimes by putting extra land, especially better land, to them. This, however, will not be done to such an extent that there will be no farms of

a suitable size for a man who has gained knowledge as a hired man and wants to start on his own account. With the greater number of cottages it becomes easier to find one vacant, either for a workman or a farmer's son or daughter who wants to get married, or for a tenant who wants to retire.[19]

The Lake District custom of the landlord owning the nucleus of the flock of Herdwick sheep and letting the flock with the farm makes it easier for young men without much capital to get a foot on the farming ladder. The reason is that the sheep's 'heafing' instinct is maintained, each ewe knowing her own area of fell and teaching her lamb to do the same, and thus the fellsides do not have to be fenced.

Managing the farms was only part of Acland's job; managing the influx of cars, caravans and campers was equally demanding. On fine weekends in the late 1950s cars ruined Tarn Hows and other beauty spots; campers covered the whole valley bottom of Great Langdale; and caravans disfigured parts of the shore of Lake Coniston. Acland realized that, if these eyesores were to go, he would have to provide alternative sites. With the co-operation of the National Park, car parks and camping sites were discreetly located and well screened, sometimes among the trees, and the results were so effective as to convince even the motoring organizations of the benefit. No restriction was placed on camping, whether at high level or on farms where there was some screening and the tenant would exercise control. The first Whitsun weekend after the Langdale campsite was opened, campers were warned on television and radio that they would no longer be allowed on the nearby Common, and staff spent the next three weekends persuading them of the advantages of sanitation and lack of litter. But not all the changes were free from dispute. On one occasion the Trust resisted the local authority's wish for more lavatories and was supported by the *Guardian*: 'the British spirit that won the

Ashes and conquered Everest surely does not require public lavatories at half-mile intervals along the Borrowdale road'.[20]

After 1961 the stream of country-house acquisitions diminished. Only six were given or transferred through the Land Fund in the remainder of the decade, but several of these had exceptionally fine gardens. Among them was Shugborough, in Staffordshire, where formal terraces lead down to the river and the Tower of the Winds and other classical monuments embellish the park. In contrast to this eighteenth-century landscape are the gardens at Anglesey Abbey near Cambridge, which had been laid out by the first Lord Fairhaven after 1926. Conceived on an unusually majestic scale for the twentieth century, the long tree-lined avenues, focused on the many statues, form a framework for the vivid colours of the dahlia garden and the picturesque corn mill on the river.

By the late 1960s the Trust, by accident rather than design, had come to own gardens dating from almost every period between the reigns of Elizabeth I and Elizabeth II. At Montacute, the Trust's second country house, the original garden walls and lay-out have survived since they were first conceived in the 1590s; at Packwood some of the topiary and the formal plan reflect the continental influences brought over by Charles II and his court at the time of the Restoration; a reaction against this formality is to be found in the larger-scale 'natural' landscape gardens surrounding the lake at Stourhead. Early in the nineteenth century formality returned at Ickworth, where vistas lead to the Capability Brown park beyond, and later the massed bedding of the Dutch garden at Lyme Park and the parterre at Cliveden reflect Victorian taste. These gardens are all designed to set off the mansions to which they relate and were transferred to the Trust by people well removed in time from those who had conceived them.

By contrast, gardens acquired in their own right were the result of the love and labour of their donors or their immediate predecessors and were given to the Trust as the best – perhaps the only – means of ensuring their survival. That was the reason why in 1961 Lieutenant-Colonel Sir Edward Bolitho gave Trengwainton in Cornwall with the magnificent magnolias and rhododendrons he had planted, and why Nigel Nicolson offered Sissinghurst Castle in Kent the following year. Sissinghurst was the work of Harold Nicolson and his wife Vita Sackville-West between 1930, when they bought the house, and 1962, when she died. 'He was the designer and she the plantsman and they worked in perfect harmony. He liked rational, classical things and she liked the poetic and romantic. . . . They wanted a strict, formal design with free, informal planting. They also wanted a garden of a private, almost secret nature.'[21] They achieved their aims by dividing the garden into separate 'rooms', each surrounded by walls or high architectural hedges, perhaps influenced by the similar concept of the earlier garden at Hidcote.

When the possibility of handing the garden to the Trust was put to Vita Sackville-West before her death, she rejected the idea violently. 'Never, never, never!' she wrote in her diary, 'not that little metal plate at my door! Nigel can do what he likes when I am dead but so long as I live, no National Trust or any other foreign body shall have my darling. No, no. I felt myself flush with rage. It is bad enough to have lost my Knole but they shan't take Sissinghurst from me!'[22]

This reaction was not due to any hostility to the Trust. Both Vita Sackville-West, as one of the original members of the Gardens Committee, and Harold Nicolson, as vice-chairman of the Executive, had great respect and affection for the Trust. But, with her deep sense of family, privacy and ownership, she could not bear the thought of losing the place she had nurtured. After her death Harold Nicolson was too old to go on alone and the decision fell to their younger son, Nigel. He explained what he felt:

My main desire and duty was to save what she and my father had created, to preserve in perpetuity the garden which, together with her books, is the legacy of her imagination. Few private gardens of this size could survive in the economic conditions of our time. It was a choice between its gradual reversion to the fields and cabbage patches from which it had emerged, and the surrender of the titular ownership to the National Trust. I chose the second alternative, without hesitation.[23]

By the time of Vita's death Sissinghurst was already celebrated, but there were differences of opinion in the Trust about whether it should be accepted. Dr George Taylor, chairman of the Gardens Committee, thought it 'not one of the great gardens of England', but Alvilde Lees-Milne, married to James Lees-Milne and a member of the Committee, disagreed: 'I and thousands of others put Sissinghurst way above such places as Sheffield Park. It is not only romantic and intimate, as well as full of interest, it also happens to have been created by a great English poet and writer. To my way of thinking it is everything a garden should be.'[24] The garden was accepted and transferred to the Trust through the Land Fund, the Historic Buildings Council helping with grants for the repair of the buildings. Nigel Nicolson still lives there.

Writing a few years later he felt that his mother would have approved what the Trust had done since her death and would have been glad to think of Sissinghurst safe and still growing. Perhaps for the reason that his mother 'thought of her garden as a book or picture, meaningless unless seen',[25] he was adamant that drastic measures should not be taken to curb the fourfold increase in visitors which had taken place between 1962 and 1970 when some members of the committees had suggested a reduction in the number of days open as well as banning young children and at least a doubling of the entrance fee. More visitors were given priority over conservation and as a result worn grass paths had to be replaced by brick or stone and other adjustments made, but none of these change the essential character of the garden.

The Trust had become the owner of the largest collection of historic gardens in the world, and their management posed a number of difficult problems. By 1954 these had become too much for the Gardens Committee to handle and it was decided to appoint a full-time adviser. The first was a qualified horticulturalist, Ellen Field, but within a few months she was severely injured in a car accident. She was followed by Graham Thomas, who served for eighteen years. Trained at the Cambridge Botanic Gardens, he had spent twenty-five years with a firm of wholesale nurserymen; he became a recognized authority on old roses, shrubs and alpine plants and wrote a number of immensely popular books, some of which he illustrated with his own drawings.

Thomas' task was not an easy one. All the head gardeners looked to him for advice, except where the donor family still managed the property and except for three properties where Lanning Roper, the eminent garden designer, took his place. Thomas answered to the Gardens Committee, which he once described as 'a dictatorial and opinionated body'. Dr George Taylor, the eminent botanist and director of Kew Gardens, succeeded Sir David Bowes-Lyon as chairman in 1961. He was not the unanimous choice. Alvilde Lees-Milne wrote to Rosse on behalf of Vita Sackville-West and herself: 'Taylor will be too busy to visit our gardens; and this I think is a great pity. . . . It worries me greatly that the Trust is accepting so many gardens just because of their interesting specimens. We ought to accept them for historic lay-out reasons.'[26] But Thomas believed that Taylor was an excellent choice. One of the troubles with the Gardens Committee at the time was that while all the members were excellent gardeners most of them drew only on their knowledge of their own gardens. Taylor's wider knowledge of plants and a range of gardens provided a useful balance.

The chain of command on important garden issues did not

lead directly from Thomas to head gardeners. An office circular issued in July 1962 instructed staff that gardens attached to historic houses were the concern of the Historic Buildings Committee and could be changed only with the approval of the regional representative, to whom the gardens adviser should report; other gardens were the responsibility of the area agent with the help of the gardens adviser, but any major changes had to be referred to the secretary. Less officially, some members of the Gardens Committee and the regional committees could not resist giving their own advice to gardeners. And the Gardens Committee might change its mind if the members present altered from one meeting to another.

Money was short as no grants were available for replanting, and gardens not attached to outstanding historic houses were not eligible for help from the Historic Buildings Council. In July 1962 Lord Crawford and Lord Wemyss, the chairman of the National Trust for Scotland, went to see ministers, but they were offered sympathy not cash. Grants were not provided for historic gardens as such until 1974, when they were introduced under the Town and Country Amenities Act, a Private Member's measure; and at about the same time the Countryside Commission began to offer funding for replanting avenues and woodlands which formed elements within a designed park.

As late as the mid-1960s, the gardens at Nostell, Ham and several other places were still disgracefully neglected; at Nunnington in Yorkshire visitors even found the gate shut and ivy climbing over the noticeboard. Some gardens had to be completely replanted, for example Hidcote, where Thomas' planting was particularly successful.

Other gardens had not been cared for since the war and were so overgrown that their earlier design had almost disappeared. The first question was often to decide the period to which the garden should be restored. Little academic help was forthcoming as the study of the history of gardens was still in its infancy (the Garden History Society was not founded until 1968). In 1969 Thomas noted:

> It happens from time to time that we have need or desire to alter our gardens. The history of our gardens is not at all well documented; most have features added by succeeding generations and it would be desirable for all concerned to know just what we are trying to preserve in each garden. For this we need to delve into archives and to examine photograph albums and search through periodicals and books. I should like to feel that we could gradually get our gardens to a safer historical footing.[27]

For the early gardens there were sometimes records of lay-out but rarely of plants. The decisions then became a matter of taste; at Montacute Vita Sackville-West used pastel shades which were replaced by stronger colours more suited to the yellow stone of the house by Mrs Reiss, the donor of Tintinhull, another Somerset garden. The Trust did not always restore gardens to their original scheme. It was largely done at Stourhead, but at Sheffield Park, also an eighteenth-century design, the Edwardian planting was kept on the ground that the rare conifers added to the exotic appearance.

Confused or contradictory instructions could lead to muddles. Writing to Gibbs, the chief agent, in January 1966, Thomas noted:

> Speaking personally I know that I have my limitations, the main one being the time available, and therefore I think it extremely important that all head gardeners should be encouraged by visits from time to time by the area agents and other members of the staff and also tenants, donors, curators, etc, but I could list a number of disastrous things that have happened in our gardens through instructions being given by people who simply cannot comprehend the full effect of what they are doing when they plant a few trees or shrubs. Examples are at Upton, Cliveden, Mount Stewart,

Ardress, Dyrham and Wightwick; these are all due to the inexperience of the donor, tenant, curator or local garden adviser. In addition we have had some disastrous happenings when local committees have looked after our gardens for the same reasons, examples are: Hidcote and Bateman's. We have also had a disaster when a head gardener assumes that he is donor, tenant and National Trust all rolled into one. I think we should all agree that a head gardener is the man to carry out the instructions of the owner of the garden. In our case we have to assume that we are the owner of each garden and try to absorb the history of the place so that we shall not make one garden like another.[28]

In his determination to avoid a uniform National Trust style, Thomas sought to develop the inherent character of each garden. One way of keeping gardens distinct was to encourage a speciality in each, particularly where there was already a worthwhile collection – hydrangeas at Trelissick, old-fashioned roses at Hidcote, rhododendron species from one valley in China at Trengwainton. Another problem he had to deal with was how to provide some interest for the six or seven months that the gardens were now open to the public. This was particularly difficult in rhododendron gardens, where there was nothing to look at during June and July. The increasing number of visitors created further difficulties. In 1966 110,000 came to Stourhead and 76,000 to Bodnant. They not only trampled grass paths to mud, they also needed such facilities as car parks and lavatories. The question of what to do about teas was a constant item on the Gardens Committee agenda.

Thomas was universally recognized to be an excellent horticulturalist, but his critics suggested that his 'eye' was sometimes at fault. Considering the demands made upon him, this was not surprising. Every garden demands a creative 'eye' and almost every horticultural problem has an aesthetic aspect. In

1969 he pressed for more help: 'I am asking for an assistant so that I shall not feel I am continually skating over problems as at present.'[29] The paragon sought would need to be of good education and be able to put over his ideas; to have an extensive knowledge of trees, shrubs and plants and their culture; to be able to use plants creatively with an artist's eye; and to have visited many famous gardens and be conversant with garden history and its changing fashions. Apart from these essential qualifications he ought to know about pests and diseases, and garden machinery, and to have spent part of his working life in a tree shrub nursery. The man who satisfied these exacting criteria was John Sales. He was appointed as Thomas' assistant in January 1971 and as his successor three years later. One of Sales' major contributions has been the recruitment of gardeners who have a greater knowledge of horticulture and planting and have done much to raise standards. Another has been the introduction of garden surveys, which have proved a model for other places.

Despite the difficulties, by the end of the decade Thomas felt that the system was working well and the head gardeners happy, 'except where we have to deal with difficult tenants or donors who often try to obstruct our endeavours while not understanding good husbandry and gardening finesse, nor the needs of gardens which are open to the public'.[30] Some years later, however, when Patrick Gibson took over the chairmanship of the Trust, he found that the chain of command was still causing confusion and initially he had doubts about the aesthetic advice received from the gardens advisers, though as time went on he became more satisfied.

Differences of opinion on the management of gardens or the number of head-office committees might have interested a gossip columnist but would never have hit the headlines. The culling of seals on the Farne Islands was another matter. When the Trust first acquired these islands, about 800 grey seals came ashore to breed, but after the 1932 Grey Seals Protection Act the numbers increased to about 3500. Some people saw this as a victory for conservation, but the Northumbrian fishermen complained that the seals were decimating the fish stocks and damaging their livelihood. On the advice of their scientists and with the approval of the Nature Conservancy, the Ministry of Agriculture asked the Trust to agree to the pups being culled with the aim of reducing numbers by a quarter. Rather unwillingly, in 1962 the Trust agreed that the cull could take place for a three-year trial period. But they had not foreseen the grisly photographs that would appear in the press or the complaints which would pour in from members. The animosity was compounded by the use of a Norwegian ship and crew who did not share the English squeamishness over shooting seals. In 1965 the Executive Committee bowed to the pressure and announced that culling would not be allowed until further research had been carried out into the ecology of the islands.

In 1968 the Natural Environment Council reported that the growing population of seals was causing malnutrition and disease for them; was damaging fisheries; and was eroding the vegetation of the islands, thus threatening the habitat of the puffins and other sea birds. Two years later the Ministry of Agriculture was given power to authorize entry on land for the purpose of killing seals. Convinced by the scientific evidence, in 1971 the Executive Committee, with the agreement of the AGM, decided that the number of seals should be reduced. On 17 August 1971, the *Daily Mirror* headlined the decision, 'Doomed for Slaughter, 3500 seals of hunger', and continued, 'Thousands of seals on a nature reserve have been sentenced to death by the National Trust'.[31] Despite the outcry, annual culls continued until 1975, when it was decided to place two wardens on one of the islands during the culling season. Unexpectedly this had the effect of dramatically reducing the number of breeding cows. No one knew where the seals had moved to until a new and thriving colony was reported in the Firth of Forth. A simple solution had been found to a painful problem.

The next public rows were set off by challenges to the Trust's right to hold land inalienably and in perpetuity unless Parliament ruled otherwise. Bath City Council, eager to attract one of the new universities, had offered a site which included 140 acres of playing fields and had undertaken that these would be relocated elsewhere. The Council set their sights on Rainbow Wood Farm, high on the hill overlooking the city from the south, which had been bequeathed to the Trust on the understanding that farming would continue. The Council proposed to provide no less than thirty-seven football pitches, together with changing rooms and a car park, but promised to preserve the beechwood which gave the farm its name and to protect outlying parts of the estate. The Trust refused to give way, arguing that there were alternative sites and that if Bath Corporation were successful potential donors would have no confidence that their gift would not in future be seized by public authorities when it suited them. This was the first time that a local authority or central government had attempted to use compulsory powers to acquire property owned by the Trust for any purpose other than roads. The case rested on the 1907 Act, which defined the Trust's objective as the permanent preservation of land and buildings of beauty and historic interest; and provided the means of securing this objective through the principle of inalienability. The Trust had followed the usual strict procedure before declaring the property inalienable: the Executive Committee had been satisfied that the land was of outstanding landscape importance, that it was of benefit to the public, and that there were no current planning objections.

Explaining the Trust's resistance to the compulsory purchase order issued by the Bath City Council, Antrim told the 1967 annual general meeting: 'We are determined to resist to the utmost, in Parliament if need be, this bureaucratic rape of land which it is our duty to protect for posterity.'[32] At the public enquiry the Trust was supported by the donor's

daughter and many citizens of Bath. Two years later the Minister of Housing and Local Government ruled in the Trust's
favour, declaring that land in the Trust's possession 'should
not be compulsorily acquired unless an overriding national
need was proved beyond any doubt and unless it was also
shown equally clearly that no reasonable alternative could be
found'.[33] The publicity had the advantage of drawing attention
to the importance of the wooded hills to the south as a back-
drop to the city and some years later when the Trust decided to
launch an appeal to protect the skyline the City Council contributed; the last piece of the jigsaw fell into place when Prior
Park was given to the Trust in 1993.

Running concurrently with Rainbow Wood was a fight at
Plymouth where the Ministry of Transport wanted to drive a
six-lane dual carriageway through Saltram Park. For the first
time in its history the Trust decided to invoke the power to
appeal to Parliament. But although Saltram Park had been
declared inalienable the Trust was on much weaker ground. At
Rainbow Wood the Trust could show that no development
proposals were contemplated at the time. At Saltram the Trust
had been aware of proposals for widening an existing road
through the park. Now a new highway was proposed. In
November 1968 the parliamentary appeal was heard by three
peers and three MPs under the chairmanship of Lord Jessel. In
his evidence Michael Trinick, the regional director, stressed
the qualities of the landscape, the difficulty of concealing the
road, and the existence of alternative routes. But the Parliamentary Committee found in favour of the government, probably because the Trust had known of the plans at the time
when the park was declared inalienable.

Battles over inalienable land multiplied when the Heath
government introduced a more ambitious roads programme
after 1970. The hardest-fought and longest-lasting controversy
concerned the proposal to construct a bypass through the Park
at Petworth, one of Capability Brown's most beautiful and

intact landscapes. The Trust launched a petition which was signed by 350,000 visitors to Trust houses, and at Petworth the route was marked out so that the public could see for themselves where the road would run. In 1975 the campaign brought a reprieve with an announcement that the public enquiry would not take place for at least a year. Nearly twenty years later the issue has still not been resolved.

Crawford had presided over the Trust during the rows about the Farne Island seals and a long-running dispute with Manchester Corporation about the extraction of water from Lakes Ullswater and Windermere. He announced his decision to retire in July 1965, a suitable time because, he thought, 'we are not at present involved in any exceptional difficulty, problem or negotiation; our coastal appeal has been launched; our work is generally making steady progress: above all . . . a successor . . . of the right age and calibre is now available'.[34] The flood of letters regretting his departure was summed up by one from the Queen Mother: 'We shall never get anyone half as good as you. Your vision, your courage and wisdom in guiding the affairs of the Trust are things that this country must be grateful for.'[35] Crawford had undoubtedly presided over the Trust with authority and aesthetic judgment during a long and formative period of its history. The reverse side of the coin, as there must always be, was that towards the end of his tenure some of his colleagues thought that he was remote and out of touch. In particular, he showed no awareness of the urgent need to reform the Trust's management or to try to deal at an early stage with the tensions building up round Commander Conrad Rawnsley, the director of Enterprise Neptune (see Chapter 10).

Not everyone was sure about Antrim, whom Crawford had chosen as his successor. Rosse wrote to Crawford, 'Ran's health often looks absolutely ghastly [he had been seriously ill in 1962] . . . and it is by no means instinctive with him to take the right decisions in such matters [aesthetics] in a way in which it certainly is for you.'[36] But Chorley approved, as did most of the Committee members: 'He has not got the Lindsay [Crawford] distinction in public, but he has enthusiasm, is ready to work hard and has excellent judgment.'[37] He was undoubtedly very different, in appearance 'a roly-poly puck with a permanent twinkle in his eye and gusts of laughter. He was shrewd and was not fooled by attempts to take advantage of his good humour,'[38] recalled Mark Norman. He was warm and direct in manner and spent much more time in the office than Crawford had done. Antrim took over at the age of fifty-four after seventeen years as chairman of the Northern Ireland Committee, nine years on the Executive and just over a year as chairman of the General Purposes Committee. In a letter to Crawford he said, 'I approach the chairmanship with humility, it is a very important and responsible one. I have a wonderful example to follow. You have been an ideal chairman and in doing that you must have made us all – mixed lot that we are – your friends.'[39]

CHAPTER TEN

Admiral Neptune
and Commander Rawnsley

ENTERPRISE NEPTUNE was not Antrim's idea, but it had his enthusiastic support. The idea came from Christopher Gibbs, the newly appointed chief agent. He had started the Pembrokeshire Coastal Appeal in the 1930s and had long been anxious to get something like it going again. He was the natural choice to represent the Trust on a high-level committee which had been appointed by Max Nicholson, the director of the Nature Conservancy Council and a member of the Trust's Estates Committee. The committee was intended to draw attention to 'the gradual and irretrievable disappearance of every part of the coastline of Britain', but Gibbs found that it 'seemed only to think of putting high-rise flats in coastal towns'.[1] He decided that it was up to the Trust to show what could be done by private enterprise and in 1962 he persuaded the Executive Committee to back his proposal for a national appeal to buy or acquire covenants over the most beautiful and unspoilt stretches of coast in England, Wales and Northern Ireland.

It was none too soon. By 1958, 16 per cent of the coast of England had been permanently built over and development had proceeded at the rate of about three and a half miles a year since the war. This figure took no account of caravans and chalets or of oil refineries, nuclear power stations or defence installations. Including these, it was estimated that by 1965 some five miles of coast were being consumed each year. The

Ministry of Housing and Local Government's survey of 1958 showed that some maritime counties had suffered more than others, the worst being Sussex with 47 per cent built up, Denbighshire with 37 per cent, Kent with 29 per cent (soon to reach 50 per cent) and Lancashire with 25 per cent. Little of the Sussex coast remained untouched apart from the Seven Sisters and Fairlight Cove, which were already owned by the Trust, and Beachy Head, which belonged to the Eastbourne Corporation.

The 1947 Town and Country Planning Act had proved less effective than had been hoped in resisting the mounting pressures from holidaymakers. A few figures explain why this had happened. In 1963, twenty-six and a half million people went on holiday in Britain, compared with fifteen million before the war and, of these, three-quarters went to the seaside. There were eight million cars on the roads and three million people stayed in caravans, which were being built at the rate of 45,000 a year, two-thirds for the home market. Yachts and pleasure craft were growing in popularity.

The Trust had a springboard for the appeal in the 175 miles of coast it already owned at the start of the campaign. It was particularly strong in Cornwall, where Michael Trinick, the regional secretary, had secured forty-three separate properties in the years 1955–60. Those which he had not persuaded the owner to give or to transfer through the Land Fund he had bought from the tail end of funds that he had spotted during his three years at head office. He had qualified as a land agent after war service in the Royal Engineers and in 1953 was appointed to Cornwall, the county from which his father's family came. It was not long before he convinced Lees-Milne that he was capable of succeeding Knollys as the historic-buildings representative and he became the first man to combine this post with that of area agent. After his interest had been aroused by the task of taking over Saltram and opening it to the public, he made himself into a well-informed local historian and, gifted with an easy pen, he found time to write a number of guidebooks. As powerful in personality as in physique, he rivalled Acland in his commitment to preserving his adopted county and in his knowledge of its people. Like Acland, the responsibility thrust on him during his youth in the army had given him an unusual authority which, before the days of regional committees and regular advice from London, he was free to operate much as he wished. He never moved from Cornwall and was therefore able to build up a network of relationships and a personal position in the county which would have been impossible had he been transferred after a period of ten years or so, which became the usual practice.

Another initiative had been taken in Northern Ireland, where in March 1962 the Ulster Coastline Appeal was launched, the year after the Giant's Causeway was acquired through the Ulster Land Fund. The latter had occasioned an altercation with the Northern Ireland Ministry of Finance, who had assumed that admission charges would be levied as they had been previously. The Regional Committee objected and was successful in establishing the principle that access to the seashore (as opposed to teashops and car parks) should always be free of charge.

Before the coastal appeal could be launched nationally, careful preparation was called for and the Trust's agents were asked to estimate how much of the coastline in their areas warranted preservation. Out of the 3083 miles of coast round England, Wales and Northern Ireland they calculated that only 900 miles were still unspoilt and of sufficient landscape value to merit saving from development. With adjoining land to protect the view and to give public access, the area they thought the Trust should try and acquire amounted to a total of 330,000 acres.

This was an immensely ambitious target and would require fund-raising on a larger scale than the Trust had ever previously attempted. A director was sought for the appeal, which was to be run separately from the Trust's existing organization. John Smith and Rathbone sifted the applications but most came from people who were either retired or lacked a record of achievement, and were not judged suitable.

Among those who were rejected was Commander Conrad Rawnsley, grandson of Canon Rawnsley, but he would not accept this rebuff: 'Do you not think that my relationship to your founder would at once appeal to the newspapers, giving them a useful peg on which to hang a story to launch your campaign?' he asked in a letter to Rathbone. 'No,' wrote Rathbone in the margin. Rawnsley followed this up with another letter: 'My bank manager writes to me today: "I do hope you get the job because I can imagine nobody more suitable for it." I am sure you would agree that whatever your terms of reference the opinion of men of such wide experience and sagacity should not be lightly put aside.'[2] His persistence and his resourcefulness when added to his 'financial dealings' persuaded John Smith that he was worth trying. Rathbone was more sceptical: 'I am not sure what these financial dealings are . . . [he had set up two businesses – an educational publishing company and a toy factory]. His experience in actual fund raising seems to have been negligible. But he is a nice and interesting man and he is very keen indeed to work for the Trust. He may be a bit of a crank.'[3]

Despite Rathbone's doubts, in February 1963 Rawnsley was appointed to start work in May. He set about his task with vigour. The name 'Enterprise Neptune' came from him, as well as the suggestion that Prince Philip should be asked to become patron and the idea that the campaign should start on St George's Day, 23 April 1965. He proposed that 500 bonfires should be lit by youth organizations, perhaps recalling that his grandfather had been responsible for the bonfires to celebrate Queen Victoria's Diamond Jubilee. So far as possible, the sites were to be those used for warning beacons at the time of the

Armada, but on this occasion instead of alerting the country to an impending attack from a seaborne enemy the fires would warn the nation of the threat from shore-based vandals, one of the objectives of the campaign being to focus attention on the issue.

Prince Philip formally launched the appeal on 11 May at a lunch given by the Lord Mayor at the Mansion House, expressing the hope 'that this operation will be seen as a turning point in our whole approach to the problems of land use' and warning that 'without some remnants of the country-side which had inspired and warmed the hearts of generations of British people, life in these islands was going to be reduced to the level of animals on a factory farm'.[4]

Crawford had already in the early 1960s taken the precaution of obtaining approval for the campaign from Sir Keith Joseph, Minister of Housing and Local Government and Minister for Welsh Affairs (1962–4), but Antrim hoped for more positive support from James Callaghan, the new Labour Chancellor of the Exchequer, who was reported to be keen to help with preserving the coast. This he did, announcing a gift of £250,000 to stimulate private donations. Together with a grant of £50,000 from the Pilgrim Trust and £64,000 raised at the Mansion House lunch, the appeal got off to an encouraging start towards the target of £2 million – Rawnsley would have preferred £5 million.

Rawnsley devoted boundless energy to the campaign, writing letters to newspapers and magazines from *The Times* to the *Women's Institute Journal*, speaking on the radio and visiting towns all over the country. Although he failed to persuade the Postmaster-General to approve a Neptune stamp or to arrange a national flag day, the campaign attracted extensive media coverage and a great many events were held. There were exhibitions of books, photographs and paintings; a colour film – *The Vanishing Coast* – was produced; forty civic functions took place, as well as private functions from balls to coffee parties, whist drives and sales of work. Many of these had been the responsibility of the campaign organization which Rawnsley had set up. By the spring of 1966 he had appointed paid regional directors in Manchester, Birmingham, Cardiff and London and a large number of volunteers, some of whom received expense allowances – over thirty county commissioners (out of sixty-three intended) as well as district commissioners, local events organizers and wardens.

To outward appearances the appeal was going well: by mid-1966 £719,000 had been raised and sixty-four properties covering sixty-six miles of coast had been acquired or were under negotiation. The first property was an area of sand burrows and saltmarsh noted for its plants and birds on the Gower Peninsula in South Wales; several other early properties were bought on the dramatic cliffs of Golden Cap in Dorset. But within the Trust there was growing disquiet at the administrative costs of the appeal and growing dissatisfaction with Rawnsley. The Neptune Committee, which had been set up in June 1964 under Antrim's chairmanship to control the campaign, had succeeded in stopping some of his more extravagant schemes, such as the purchase of the *Medway Queen* to sail round the coastal resorts, but Rawnsley had looked to them for assistance rather than direction and had never fully accepted their authority nor that of Rathbone.

The underlying personality problem was already apparent in a letter Rawnsley wrote to Crawford a few weeks after he started work: 'I have never been in any doubt about my ability to do the job . . . provided that certain conditions essential to the success of any enterprise exist. These are firstly that direction from above shall be clear and unwavering and secondly that guidance shall not degenerate into interference.'[5] This was followed by a letter to Antrim complaining that he did not enjoy the confidence of Crawford or Rathbone and that he encountered hostility from a large section of the Executive.

By the end of 1963 Antrim was sufficiently concerned to tell

Rathbone that if Crawford and the rest of the Committee were not in favour of the appeal it ought to be stopped at once. Rathbone wrote to reassure him: 'I think Crawford, Lady Dalton, Kenneth Robinson [Labour MP for St Pancras North] and Michael Rosse, too, are not so much against the appeal as worried by Rawnsley and his grandiose ideas. These have already been much modified. John [Smith] agrees with me that it would be a great shame to get rid of Rawnsley. . . . Ruby [Holland-Martin] agrees provided he is controlled. I believe the appeal is a good thing for the Trust.'[6] Rawnsley continued writing letters of complaint, even to Crawford, causing Antrim to send a prophetic note to Rathbone: 'It confirms to me that Rawnsley is a bit off his head and he certainly has no judgment of either people or circumstances. Let's hope he does no harm.'[7]

As time went on Rawnsley made less effort to disguise his view that 'the Trust is one of the surviving institutions of feudalism and a protégé of the old landowning class'.[8] Early in 1965 in a conversation with Rathbone he referred to 'the belted earls of the Trust' (a remark that was quickly passed on). A few months later Antrim, one of the 'belted earls', received a letter beginning 'I really must protest about your performance this afternoon' (after Antrim complained in public about a letter in which the name of the addressee was wrongly spelt) and continuing 'anyone who is worried about an occasional misspelling of his name is not worth bothering about anyway for he can neither be a gentleman nor have any sense of proportion'.

On 9 July 1965 Rawnsley wrote to Richard Dimbleby, the celebrated television master of ceremonies, 'I believe you already have taken the point that Neptune is really and truly for the man in the street and so is the National Trust. It is a thousand pities that this has not been the public image of the Trust since before the war when the stately-home scheme bedevilled it but I'm sure Neptune will in time redress the damage.'⁹ Rathbone complained, but Rawnsley was unrepentant:

> Although I have embraced the principle that current decisions by committees should be upheld, whether one thinks them right or wrong, it is to my mind questionable whether this position even is really tenable today. . . . The stifling of criticism of any institution whatever by any person and in any circumstance whatever is undemocratic and therefore cannot be right except in the face of enemy in wartime. I can only suggest that you move in too closed a circle of NT supporters to sense the temper of public opinion and this is understandable since you have made it your life for fifteen years.¹⁰

Faced with this reply the General Purposes Committee directed Rathbone to tell Rawnsley at once 'that, unless you carry out to the full the instructions contained in my memorandum of 14th July and stop criticizing to outside people the policy of the Executive Committee, your employment with the Trust will have to be terminated'.¹¹

Rawnsley's activities landed the Trust in a number of other embarrassing incidents. They included a public attack on the Birmingham City Council for their lack of support; an acrimonious quarrel about a piece of land owned by Lord Egremont between Rawnsley's cottage and the park wall at Petworth; and a letter to 'Peterborough' of the *Daily Telegraph*, telling him of a plan, before the directors had consented to it, for Sotheby's to auction, free of commission, some important objects.

Rawnsley was particularly tactless in Wales. Not only did he upset Colonel Wynne-Finch, chairman of the Committee for Wales, by a letter in which he referred to the Welsh as 'the tribes', he also upset his own commissioner for Monmouthshire, Lieutenant-Colonel W. R. Crawshay, by appealing for funds from the county council himself when it had been agreed that all appeals should come from within the Principality. Crawshay wrote to Wynne-Finch to say that Rawnsley had made the most disastrous impression in Wales and added, 'We wish to see no part of him in South Wales again.'[12]

Disagreements extended beyond questions of language and behaviour to important aspects of policy. One issue concerned the balance between conservation and access. Rawnsley pressed for more caravan sites and more facilities for the public in general in the belief that if Neptune was to be successful, and indeed if the Trust was to survive, the image that 'it shuts things up' must be changed. Rathbone observed that camping and caravans had been allowed for years provided they were unobtrusive and concluded that Rawnsley was under the delusion that he had revolutionized the Trust's policy when in fact nothing had changed – full access had always been given to all open spaces subject to the needs of agriculture and forestry.

Another issue concerned land acquisition. Rawnsley wanted coastal landowners to recognize their responsibility for giving public access and, if necessary, be forced to sell. Crawford strongly disagreed: 'We must not give the impression to anyone that we are applying pressure for the purchase of coastland. I am satisfied providing Rawnsley does what he is told that our policy is clear, settled and unchanged. We propose to buy coastline out of money raised through the Neptune enterprise only from willing sellers.'[13]

Rathbone remained remarkably forbearing in the face of Rawnsley's refusal to obey instructions, though he did say he wondered how Rawnsley had stood up to naval discipline when he was unable to fit in with an organization as loose as the Trust. He was prepared to endure the friction so long as he thought that Rawnsley would be able to raise £2 million for Enterprise Neptune. But as the cost of the campaign mounted he became worried that Rawnsley was losing touch with reality and living in a world of fantasy. He lost faith that the private army of directors, commissioners and organizers could rouse public opinion so as to attract a vast number of 'widow's mites', as this idea was seen to be based more on hope that on any real evidence. The appeal to local authorities and business could clearly have been conducted more economically through the Trust's own staff. The fact that the situation had reached this point was an indication of the weakness of the Trust's overall direction: it was unwise to set up a separate Neptune directorate, then to entrust it to a man with no experience of fund-raising, and having done so to allow Rawnsley to operate for years when confidence in his judgment and management ability was lacking.

In the spring of 1966 the Neptune Committee reviewed the prospects of the Neptune campaign in the light of the current economic situation – a 'stop' to the economy and a freeze on salaries and wages. On the Committee were the Trust's emi-nent businessmen: Ruby Holland-Martin, the honorary treasurer; Mark Norman, managing director of Lazard Brothers; and Pat Gibson, chairman of Pearson Longman. The Committee reached the view that the £2 million target could not be achieved by the end of 1967, that Neptune should be regarded as a long-term campaign and that costs must be reduced.

On 14 October the Executive decided to wind up the separ-ate Neptune Directorate and incorporate it in the general organization from 31 March 1967. This was obviously a con-venient opportunity to get rid of Rawnsley, and on 17 October Antrim informed him that he was to be given six months' notice, but that he would then be treated generously. Rawnsley was understandably upset. He was confident that if the campaign were to be carried on until May 1968, as origin-ally planned, it would continue to draw in substantial funds. He felt that he had been badly treated: he had given up his position as managing director of his own firm; at the age of sixty he would have difficulty in finding another job and, not qualifying for a pension from the Trust, he would be in financial difficulties.

A press conference had already been arranged for 25 October at which Rawnsley was to meet local journalists at Saltram, on the edge of Plymouth. He arrived late, looking ill and strained, and launched into a bitter attack on the Trust, accusing it of being 'bankrupt in ideas, bankrupt in leadership, bankrupt in the common touch, bankrupt in [its] sense of what the people need and in the alacrity with which [it] set about providing it. Where are the working-class members of the Trust, and the youth? – they don't exist.' He continued:

> I had heard and read so often – in the Trust's own publicity – what a good job it was doing. I looked at the job perhaps with a too critical eye. It was, after all, the child of my grandfather's imagination.... Where was the fire and enthusiasm of those early days? It seemed to me to have become a part of the establishment, an inert and amorphous organization proceeding by the sheer momentum given to it by those who continued to bequeath their wealth and property to it, as often as not to escape death duties.

Rawnsley described an institution 'run on the old-boy net' and asked, 'Where was the democracy in this? Where was the life and vigour which a proper system of representation gives to an organization?'[14] The speech was telephoned through to the London office, and Rawnsley was immediately suspended; next morning his attack made front-page news. It was the beginning of the most painful row in the Trust's history and was to lead to far-reaching changes.

The annual general meeting was due to take place in Cheltenham on 12 November. In the three weeks which elapsed Rawnsley widened his attack to challenge the constitution and policy of the Trust. In an article published by the *Observer* headed 'No dogs, no picnics, hardly any caravans', he asked what the Trust protected its land and buildings for. Why were houses not used for dances and other activities and its land for leisure pursuits? Was it right for the land to be hunted over and people confined to paths because the shooting rights were let? And he asked whether the volunteers recruited for Neptune could not form the framework of a democratic system of representation and government.[15]

He was supported by John Grigg in the *Guardian*. Grigg welcomed the public discussion on the grounds that 'some of those who run the Trust seem more interested in preserving places *from* the public than *for* it, and that while there was something to be said for maintaining the traditional links of the Trust's houses with particular families there was nothing at all to be said for allowing the Trust to seem a country gentleman's benevolent society'.[16]

The Pump Room at Cheltenham had been booked for the annual general meeting and would have been large enough for the normal attendance. But many more members turned up than ever before and the doors had to be closed while members were still clamouring to get in. They had to be content with looking through the windows and listening to a loudspeaker extension. Antrim did his best to defend the Trust's policies against what he described as 'the increasingly violent and indiscriminate attacks by Commander Rawnsley', adding, 'It is impossible to believe that his criticisms, however worthy some of them may be of discussion, are not motivated very largely by a feeling of considerable grievance and spite.' Of Neptune he said:

> the campaign has had considerable success. It has raised £831,155 . . . but it has cost £116,300 to raise money . . . about 20 per cent if we subtract the £300,000 from the government and the Pilgrim Trust. Your Executive Committee came to the conclusion . . . for many reasons, including the financial climate of the second half of this year, that it was probable that the cost of raising money for Neptune would increase

rather than go down. It had therefore been decided that after March 1967 the campaign would be continued as an integrated part of the Trust at a smaller cost.

The main point he conceded was that 'we ought to bend our minds to the problem of how to ensure that we continue to get the expert representation that we have now [on the committees] combined with a further degree of election from among the members of the Trust'.[17] One observer felt that the meeting 'showed a distinct coolness towards the platform'. Rawnsley replied, brilliantly at first, but then with less effect as he devoted too much time to his personal grievance. He claimed that the decision to reduce costs was untimely and too drastic and defended himself for criticizing the Trust in public on the grounds that he had 'had to decide between loyalty to the Trust as an institution and loyalty to its ideals'.[18] He then asked members to sign a requisition for an extraordinary general meeting to discuss the administration of the Trust and its attitude to its members and the public. This proposal, which needed only thirty signatures, was carried. The Central Hall, Westminster was booked and the date was set for 11 February 1967. Antrim's speech had not gone far enough to satisfy the critics, including John Grigg, who in another *Guardian* article concluded, 'the hints of change were too limited in scope and too grudging in tone to give any assurance that what is wrong with the Trust will be put right unless there is sustained pressure.'[19]

Rawnsley set about organizing a Members' Movement for the Reform of the National Trust with the same urgency and zeal that he had displayed in starting Enterprise Neptune. He quickly recruited a number of sympathizers. Among those who attended a meeting of the 'requisitionists' as they called themselves were Robert Mowat, a chartered accountant from Cheshire, who was criticial of the presentation of the accounts in the annual reports; John Grigg; Mrs Mary Moorman, a daughter of G. M. Trevelyan and wife of the Bishop of Ripon; and several people who had personal grievances against the Trust. Raymond Cochrane, who became chairman of the reform group and was its main source of funds, owned the Cotswold village of Guiting Power which some years earlier he had offered to the Trust. This offer had been turned down on the advice of Nicholas Ridley, the local MP and a member of the Executive Committee, who had argued that as Cochrane was known to be rich the village was in no danger. But he failed to foresee the day when unspoilt villages would be as rare as country houses with their contents.

Rawnsley wrote a statement entitled 'Case for Reform of the National Trust' with the help of some advice from John Grigg. In it he developed a number of his earlier lines of attack, the most telling being those on the constitution, public access and Enterprise Neptune. He claimed that the Council was ineffective, the Executive Committee unrepresentative and the elections undemocratic, and that this produced 'a ruling hierarchy which lacks the common touch and is remote from the people'. He criticized the bewildering opening arrangements for the historic houses ('as easy to be clear about as a continental railway timetable', commented *The Times*) and suggested that 'the Trust will forfeit all pretence of honour if it fails to live up to its duty out of an exaggerated sense of obligation to the heirs or assigns of donors'.[20] This was in reply to Antrim's remark at the AGM that 'when we received these gifts we made promises to the donors and we should forfeit all pretence of honour if we failed to live up to those promises'.[21] Rawnsley linked the question of access to a claim that donors and their heirs 'escaped heavy estate duty and can live rent and repairs free for ever'. Turning to open spaces, Rawnsley, who had kept Bob Latham, the Trust's solicitor, occupied for a whole year in supplying him with facts and figures, cited the number of acres where access was restricted and the number let for shooting and again argued that there were not enough camping sites for young people or car parks for their seniors. On Enterprise Neptune he gave a string of statistics and forecast that the economies would bring about the appeal's collapse.

The Trust circulated Rawnsley's memorandum to members together with a short response. This did not attempt to rebut Rawnsley's criticisms of the constitution, other than by listing the eminent professionals who served on its committees. But it did reply to the criticisms about access, stating that at only a 'small minority' of houses was it inadequate and that opening was continually being improved, and arguing that 'there are cases where without some compromise on the number of opening days, especially during the lifetime of the donor, houses would be lost to the nation. The limited nature of the opening arrangements is temporary; the preservation of the house is permanent.'[22] It was pointed out that the family occupation represented a substantial saving to the Trust because donors acted as unpaid caretakers and made large contributions to rates and heating which would otherwise fall on the Trust. The right of some donor families to live rent free was not mentioned.

The Trust was less defensive on the question of access to the countryside: 'The Trust's first duty is conservation for all time. . . . [Its] obligation is to hand on to posterity a fragment . . . of what England was once like.'[23] Facilities must come second to preservation, but would continue to be provided wherever they did not conflict with the Trust's primary objective. On Neptune the Trust repeated its undertaking that the objective would be vigorously pursued by a permanent staff supported by voluntary helpers.

Both sides made elaborate preparations for the EGM. Rawnsley placed advertisements in *The Times* and the provincial press asking people who had criticisms of the Trust's administration or property management to write to him. These produced a number of letters about badly managed farms, entry to houses seen as 'a favour and not a right' and lack of campsites, as well as one complaining that the coast was being preserved 'almost solely for the seagulls'. Many of the letters applauded Rawnsley's fight against the undemocratic way the Trust was run.

On 18 January, the Trust held a press conference in the morning and Rawnsley another in the afternoon at which he gave notice that he would put two resolutions to the EGM. The first was that an independent Committee of Enquiry of members should be appointed to consider all aspects of the Trust's constitution, policy, finance and administration, and the second that Enterprise Neptune should be vigorously and resolutely prosecuted at full strength until May 1968. He announced that if the resolutions were not carried the reform group would propose that they should be submitted to a postal vote of all members. Rawnsley summarized the group's purpose as 'to ensure a more attentive regard by the Trust for the purposes laid down in its Acts, and the adoption of a more liberal outlook towards the leisure needs of the people. . . . These changes will not, we believe, be achieved unless the constitution is amended so that ordinary men and women in the membership may participate in the government of the Trust.'[24] At his morning press conference Antrim, taking an attitude of 'no appeasement', opposed the enquiry: 'We are reforming all the time. . . . We do not believe there is any need for it. We are confident that we are running the Trust in the interests of the public and sufficiently well not to bother with that sort of thing.' The Trust might have its faults but 'nothing major'.[25]

The *Daily Telegraph* fully supported Antrim, but *The Times* felt that Rawnsley had made some palpable hits, especially on the constitution and restricted opening of the houses. The *Guardian*, whose columnist John Grigg continued to back Rawnsley, said that 'Commander Rawnsley's reformers do not inspire much confidence. The Trust at present is thoroughly oligarchic but extremely expert. The reformers would be in danger of substituting properly elected but faceless and quite probably inexpert committee-men.'[26]

Acland, always a keen intelligence agent, managed to read a copy of the speech that Rawnsley had prepared for the EGM. He reported that 'he has been clever enough to pick up the justifiable criticisms as well as the others. You are going to need a very good speech from the floor. Your counter publicity should (a) show that he was really making a takeover bid for the Trust and (b) plug the point that his own Neptune people had passed a resolution denying him. As far as Neptune is concerned he must be made to look like a one-man band.'[27]

The Trust had also been busy: a tactical sub-committee had been appointed to direct plans for the EGM and Rathbone had written some thirty letters a day, mostly urging members to come and give their support.

Len Clark, who represented the Youth Hostels Association on the Council, was anxious that it should intervene in the crisis. This was the first, and perhaps the last, time that any nominating organization had attempted to play such an influential role. Clark consulted his principal colleagues, several of whom were well acquainted with the Trust through membership of a regional committee or a National Park planning board. They agreed that, while they did not wish to be associated with Rawnsley, some reforms were needed if the Trust were not to find itself in the position of the medieval Church at the time of the Reformation. Clark's first move was to persuade Rathbone that the Council's proposed motion expressing complete satisfaction with the government and policies of the Trust was far too smug. After almost daily exchanges the motion was amended to read, 'That, recognizing the achievements of the Trust, this meeting gives its full support to the Council and is satisfied that the Council will give due consideration to the views expressed by members at this meeting'.[28] Clark next put down a motion in the name of the YHA expressing confidence in the Trust's work but urging that contacts with members of the public should be improved. This, he hoped, would help to give members the opportunity to participate in the debate and avoid the meeting being polarized between the two extremes.

On Saturday, 11 February 1967, some 3500 members crowded into the Central Hall, Westminster, a few of whom had to be relegated to the overflow rooms. Anticipating that the meeting might get out of hand, Trinick and some of the strong young agents were sitting in the front row ready to throw out troublemakers. To their disappointment their services were not required, although the meeting lasted for five hours and once or twice broke into uproar. Antrim conducted it firmly but fairly, giving the opposition plenty of time to be heard. His opening remarks set the tone: 'There are no rules laid down for this meeting. You must accept my rulings otherwise we shall spend all day wrangling over procedure.' He then revealed that it had been necessary to reword two of the resolutions by changing their form from an instruction to a recommendation, for the reason that 'an extraordinary meeting cannot tell the Council how to conduct its business. In the Act of Parliament the Council is the only body that can decide the policy of the National Trust.'[29] A number of hecklers immediately stood up with shouts of 'Disgraceful!', 'Why was it on the agenda then?', 'Have we no rights as members?'

Neither side had chosen their opening speakers very skilfully. The first resolution was proposed on behalf of the reform group by Mowat, but he soon lost the sympathy of the audience: his 'angry restatement of his leader's indignation with the Council had nothing to do with the case, and his attempt to use impassioned oratory where a quiet and reasonable argument was needed was unfortunate'. Chorley, who replied on behalf of the Council, was equally unpersuasive. One witness remembers, 'Around me there was a strong anti-Rawnsley crowd but they were not anti-reform. While Chorley had several good points which he made rather haphazardly, I didn't feel the case for the Council had been convincingly and clearly made.'[30] Chorley added the surprising statement that 'You cannot run an organization of 165,000 members on a democratic system.' Later in the debate some speakers for the Council answered specific criticisms more effectively, the two most applauded being those defending the Trust's work in preserving 'the peace and beauty of the countryside that is Britain'. The reform group's resolutions were defeated, and the Council's resolution, moved in poetic style by John Betjeman and in parliamentary style by Lord Radcliffe, was carried, as was the YHA resolution moved by Len Clark. Rawnsley was able to force the Council to ballot all 165,000 members, as for this he needed only twenty votes. Despite the preparations for the meeting, the Trust had not arranged for a count and after the meeting there was some dispute about the number of members voting for and against the various resolutions in the main and subsidiary halls. Replying to a complaint, Antrim wrote to say: 'Two expert birdwatchers on the platform, accustomed to judging large bodies of birds in flight, put the figures at about 5–2 against the first resolution.'[31] Perhaps the most representative reaction to the day's proceedings was expressed by a woman as she left: 'A shake-up would be a good thing but Commander Rawnsley has not gone the right way about it.'[32] Clark's feeling was that, although the rebellion had failed, there was considerable distrust of the leadership.

Antrim and Rathbone awaited the result of the postal vote with some anxiety. In the event they won by 35,000 to 15,000. But 15,000 was an uncomfortable minority. It was clear that reforms were needed, as Lord Bridges, who had become chairman of the Headley Heath local committee after retiring from the Treasury, acknowledged in a letter to *The Times*.

In April the Council decided to set up a Committee of Enquiry composed of three of their own members with a distinguished outsider as chairman. The first choice for chairman, Lord Robbins, declined, and the Executive then decided that Sir Henry Benson, the second choice, should be invited to serve as a member, but that the chairman should be Antrim. Both Sir William Hayter and Len Clark, who had already agreed to join the Committee, strongly disagreed. As Hayter told Clark, 'I personally have complete confidence in Lord Antrim, but I am sure that his nomination as chairman of the Committee would not inspire confidence and might even excite derision, and indeed I am very doubtful whether I could agree to serve on a committee constituted in this way.'[33] Antrim was soon persuaded, and in July the Council agreed to Sir Henry Benson as chairman, with wide terms of reference: 'to review the management, organization and responsibilities of the Trust and to consider possible amendment to the National Trust Act'.

Most of the reform group were satisfied that they had achieved their immediate objective. At the next AGM in November 1968 Cochrane, their chairman, announced that they would suspend all collective activity until the Committee's recommendations were made public, a decision rendered easier by the fact that a hip operation had put Rawnsley out of action for several months. The AGM was attended by a record number of members, but no hostile resolutions were moved and only a few questions about accounts and publicity were asked. Lord Goodman, chairman of the Arts Council, kindly but over-reassuringly told members that they 'need not have the slightest worry that the Trust's image has suffered'.

Whatever the state of the Trust's image, the Neptune Appeal was doing well. Antrim was able to announce that it had passed the £1 million mark by mid-1967, that the receipts exceeded those received from the previous year by £44,000, that the costs had fallen from £60,000 to £25,000 and that the length of coast purchased through Neptune amounted to seventy-five miles (146,000 acres), with a further thirty-seven

miles under negotiation. The campaign had also been successful in alerting the authorities to the despoliation of the coast: in
1970 the newly established Countryside Commission published a report identifying thirty-four separate stretches of
coast in England and Wales as 'Heritage Coasts' which should
have a special claim to protection.

After the row had died down, the press and at least a few
members of the Trust's Council came to the conclusion that
Rawnsley had done the Trust a considerable service even
though by the manner of doing it he had damaged himself and
had left deep personal scars on his colleagues. Some believed
that he had risked destroying the Trust and had helped to
precipitate Rathbone's early retirement. Rawnsley felt that he
had been ungenerously treated and was supported in this by
John Smith, who in a letter to Antrim urged, without success,
that something should be done for him since 'his notice was
presumably not made conditional on his remaining silent'.[34]
Critics were bitterly resented by officers of the Trust and, as
after Suez, there were friends who did not speak to each other
for several years.

CHAPTER ELEVEN

Trial by Benson

THE BENSON Committee completed their report eighteen months after their appointment in July 1967 and made a number of far-reaching proposals for modernizing the constitution, devolving management to the regions, introducing stringent financial controls and improving access for the public. Prompted perhaps by moves to persuade the government to order an independent inquiry, Antrim did not delay: the Committee's report was discussed at a series of meetings of the General Purposes Committee, of the Council and of ordinary members before being submitted to the next annual general meeting and incorporated in a new Act of Parliament. The principal reforms were put in hand and were implemented within a few years.

Sir Henry Benson, the chairman of the Committee of Enquiry, was the leading chartered accountant of his day. Born in South Africa, he was a partner in Cooper Brothers (later Coopers and Lybrand) for more than forty years and was made a life peer in 1981. He liked work which transcended commercial accountancy and was just finishing a report into the financial structure of the racing industry when he was approached by the Trust. A large, genial man, accustomed to brushing aside opposition, he worked at great speed and wrote the report in longhand himself. 'Of all the tasks I have undertaken,' he wrote later, 'I think results of this Committee's work have given me the greatest satisfaction.'[1]

He was joined on the committee by Patrick Gibson, who was then a director of the *Financial Times* among other companies and was to succeed Antrim as chairman of the Trust, and by Sir William Hayter, formerly Ambassador to Moscow and at that time Warden of New College, Oxford. Both of them contributed a number of the initiatives proposed in the report, as did the fourth member Len Clark. He had been on the Trust's Council since 1962 and during twenty-nine years' service became one of its most influential members. Born in 1916 in Islington, he left Highbury Grammar School at the age of sixteen to become a clerk in the service of the London County Council, where he worked in a number of departments until, as the London Ambulance Administrator, he joined the National Health Service. He served on the Properties and Executive Committees of the Trust for twenty-three years and was chairman of the Southern Regional Committee between 1978 and 1986. Even after his retirement, his instinct for getting to the heart of a problem and his wise advice, laced with witty comments, continued to be much in demand. When Lord Oliver was commissioned to study the workings of the constitution over twenty years after the Benson report, Clark became one of the two assessors.

The Benson Committee could hardly have been described as an invasion of the Visigoths into the Trust. In the words of Rathbone's successor, Sir John Winnifrith, 'It was a cosy family party. All three members were on the Trust's Council and even Sir Henry was one of the Trust's shooting tenants.'[2]* Jack Boles, later director-general, was one of the two secretaries. But Benson was carefully impartial when interviewing witnesses. Cochrane found him 'not only efficient but sympathetic – I think genuinely so – and open-minded. He also has a sense of humour, which to my mind is sadly lacking at Q.A.G. [Queen Anne's Gate].'[3]

The Committee invited members, staff and organizations

---

*The Drovers Estate, Sussex, had been left to the Trust in 1942 by Benson's uncle Sir Francis D'Arcy Cooper with the proviso that he and a cousin should enjoy the shooting rights during their lifetime.

connected with the Trust to submit their views in writing and interviewed thirty-three people individually. Among them were the leading critics, including Rawnsley, whose claim that the Trust was run by homosexuals was not calculated to create a favourable impression. Benson recorded that 'At the end of a long and tiresome meeting I said spontaneously and with conviction that I was sorry I could not help him because I had never been able to detect a homosexual from a heterosexual.'[4] The Committee's starting point was what they termed 'an internal dispute over the purposes, policies, and management of the Trust' which they thought had been needlessly damaging to the Trust's image but had pointed to the need for a review of some of the policies and procedures hitherto taken for granted.

For Benson the most pressing problem was the Trust's financial position, but for others it was the outdated constitution, which had not been touched since it was enacted in 1907 for a few hundred supporters, many of whom could be expected to attend an annual general meeting in person. With a membership of 165,000 this was no longer possible. The question was how to enable members to participate in running the Trust. Benson's solution was that they should continue to have the right to elect half the Council and to express their views on resolutions and that they should in future be able to vote either by proxy or in person. The alternative of postal voting was turned down because 'it often causes members to cast their votes [before the debate at an AGM] on matters about which they know little or nothing. In the case of a proxy vote, however, members are able to rely on the judgment of the persons whom they appoint as proxy and who can be expected to know what is likely to be in the best interests of the Trust.'[5] The Committee proposed that the final authority should continue to rest with the Council rather than with the members: 'The members are entitled to make their views known and it is most unlikely that a view which clearly reflects the opinion of a majority of members will be disregarded. But the ultimate

Skiddaw reflected in Derwentwater, Lake District

View of Winnats Pass from Mam Tor, Peak District

Wicken Fen, Cambridgeshire: reed cutting in January

Luccombe, Holnicote Estate, Somerset

Stourhead, Wiltshire: temple of Apollo in May

Hidcote Manor Garden, Gloucestershire, shortly after restoration by Graham Thomas, *c.*1967

Long Gallery, Blickling Hall,
Norfolk

Sudbury Hall, Derbyshire: staircase after
redecoration by John Fowler, 1969

Cragside, Northumberland

Cliffs at Kynance Cove, Cornwall

View from Plas Newydd, Anglesey, across the Menai Strait
to Snowdonia

Early morning over the Stone Circle, Avebury, Wiltshire

View of Fountains Abbey from Studley Royal garden, North Yorkshire

Quarry Bank cotton mill, Styal, Cheshire

Souter Lighthouse, South Shields, Tyne and Wear

Uppark, West Sussex: light, from fire engines, shining on the burning house, 30 August 1989. *Inset:* after restoration, 1993.

Kingston Lacy, Dorset, after the storm, 25 January 1990

decisions of policy must be taken by the Council.' Half the members of the Council would continue to be nominated so as to give 'prestige, strength and stability to the Council and to provide a close association with . . . organizations' relevant to its work. This combination of election and nomination has, over a hundred years, thrown up able and committed people, from a variety of backgrounds and with a wide range of experience.

The emphasis on the role of the Council meant that its meetings would have to be taken seriously instead of being regarded purely as a formality. Len Clark had summed up the existing practice in a letter he had sent to Antrim a few years before in January 1966. He had served for four years and had attended every meeting except one: 'It is doubtful', he said,

> whether the aggregate of [my] attendance time has exceeded one hour. . . . In most organizations, of course, membership of the Council carries with it some committee responsibilities. I realize that in the case of the Trust committees are filled by the method of more circumspect selection. Mr Rathbone has kindly sent me copies of the Executive Committee minutes, but otherwise I must confess that my main source of information about the Trust comes from the press – it may be that Council members are expected to make themselves active on the Trust's behalf in other directions. If not it would seem a pity that they are not more closely involved in the day-to-day work of the Trust, unless the Council is regarded as only a decorative part of the organization.

Antrim had replied that 'As the Council delegates most of its work to the Executive Committee, I do not see how, without duplication, it would really be practicable for them to do any more than they do at present.'[6] Benson tried to answer this question, suggesting that the Council's responsibilities should include such matters as criteria for new properties, arrangements for public access and standards of maintenance; in addition, it should receive a quarterly report on the Trust's affairs.

But the exact role of a more active Council proved difficult to distinguish from that of the Executive Committee, and fifteen years later the problem was referred to another committee chaired by Simon Hornby.

At long last the jungle of head-office committees was tackled. Under the Executive Committee there would be only two – the Finance Committee and the Properties Committee, which would combine the Estates and Historic Buildings Committee and would be chaired by Rosse. This amalgamation was not supported by most of the latter Committee, who believed that aesthetic matters 'should be tackled by a group of people entirely devoted to questions of taste, qualified to judge works of art, and with expert knowledge and experience. We do not believe that a "balanced" committee of Historic Buildings people and Estates people will settle such matters anything like so well. And of course we believe that the "aesthetic" aspect of the Trust's work is of absolutely primary importance.'[7] But they were overruled. Some years later Lees-Milne complained that the Properties Committee was 'far too big and not the fun it used to be. I get very bored listening to Major Hadden Paton discoursing on fowl pest, parsnips, root rape and all those disgusting vegetables.'[8] Rosse was perhaps the only member of the Trust's hierarchy who could have welded together the Estates and Historic Buildings Committees. His natural authority and his knowledge of such different matters as running an estate, propagating a magnolia or passing judgment on a painting enabled him to unite aesthetes, gardeners and foresters into a single committee. However, the views of the archaeologists, ecologists and other eminent professionals who had served on the separate committees were much missed, and within a few years they were reassembled in a more orderly mechanism of panels advising the Properties Committee.

In 1969 Mark Norman was appointed chairman of the Finance Committee and he came to play as influential a role in the Trust as had his father Ronnie Norman a generation

earlier. He had already been a member of the General Purposes Committee for fifteen years and he was also on the
Council and the Executive Committee; he later became deputy
chairman. Another Etonian, he had gone on to Magdalen
College, Oxford. Professionally he emulated his uncle Montagu Norman more than his father, spending most of his
working life in the City as a partner in Edward de Stein and
then as managing director of Lazard Brothers. His gusto,
charisma and humour made the Finance Committee more fun
than any of the other committees. Presiding over the finances
of the Trust at a time when not all the chief finance officers
were of the calibre necessary to control what had become a
complex business with a very large landholding, he tackled the
formidable financial problems that Benson had identified. To
help him on the Finance Committee he recruited one of Benson's young partners. This was Roger Chorley (the son of Theo
Chorley), who in 1991 became chairman of the Trust. It was he
who devised the formula which was introduced to calculate
the endowment needed to avoid the large deficits which had
occurred even on houses accepted with what at the time were
thought to be generous sums. This has proved of critical
importance in the takeover of properties with a very large
deficit.

One of the few major Benson recommendations to be
turned down by the Council was that the chairmen of regional
committees should not be members of the Executive Committee, for the reason that 'they would often be in an invidious
position, because they would tend to emphasize the special
needs of their own regions'. Norman disagreed with this view
and carried the Council with him. He thought it dangerous to
let loose the regional chairmen on policy and finance without
experiencing the constraints imposed by membership of the
Executive Committee, where such matters were settled. This
arrangement prevented a rebellion from any of the regional
chairmen, but it made the Executive Committee much too big.

With the abolition of the General Purposes Committee there was no longer a small group meeting at monthly intervals to direct the affairs of the Trust; *ad hoc* meetings between the chairman of the Trust, the vice-chairman and the chairmen of the Properties and Finance Committees to deal with urgent business were not an effective substitute. The balance of power in the Trust began to shift from the voluntary chairmen to the full-time executives, and the move was accentuated as the staff became increasingly professional.

Benson's second far-reaching recommendation was for the management of the Trust to be devolved to fourteen regional committees, subject to the control of finance and new acquisitions being retained by head office. Their members were to be appointed by the Executive Committee and their chairmen approved by the Council. The whole question of the regions excited more discussion and disagreement than did any other issue. The most extreme decentralist position was taken by Nicholas Ridley, Conservative MP and later a member of Margaret Thatcher's Cabinet. He thought that regional committees should run their own affairs without any interference and 'should evolve their own form of local democracy'.[9] Janet Upcott, the last remaining link with Octavia Hill and still a member of the Estates Committee, was also in favour 'of enhancing the representative character of the regional committees', but she emphasized that national standards had to be maintained. She concluded, 'I should now rejoice if a modernizing movement fulfilled the hopes of Octavia Hill, whose deepest interest was in the enjoyment by the common people of the pleasure offered by the National Trust.'[10] Len Clark explained why the Benson Committee had rejected the idea of electing members of regional committees:

> We spent quite some time considering the best methods of appointing regional committees. The method of elections through an annual general meeting was obviously one, especially when considered against the current movement

for more member participation in the affairs of the Trust. In common with my colleagues I came down finally against this method. . . . the best committee is one whose members will be effectively in touch with the management and running of the Trust properties in the area. Open-style elections are liable to throw up the loqacious and litigious in an organization such as ours.[11]

The election of two members had in fact been tried in Northern Ireland when Antrim was chairman of the Committee, but it did not work well for the reasons suggested by Len Clark.

At the time the Benson Report was published there were eight regional committees. Their chairmen, although by no means carbon copies of each other, were, with one exception, all from a country magnate and/or country gentleman background, five of them Etonians. A long-standing member of the Executive Committee recorded notes on several: Robin Herbert for Wales ('tall, aloof Welsh landowner, a top gardener, very able and ambitious'), Major-General Sir Julian Gascoigne for Devon and Cornwall ('vociferously objected to any impediment from Q.A.G. . . . and steamed out of meetings if thwarted') and Sir Marcus Worsley for Yorkshire ('a hereditary Yorkshire landowner of great authority'). There were also the Earl of Clanwilliam for Northern Ireland, the Earl of Euston, who had not then succeeded as Duke of Grafton, for East Anglia, Admiral Sir Deric Holland-Martin for Severn, Peter Scott for the Lake District and Major Peter Orde for Northumberland and Durham. A year later they were joined by Anthony Head (also an old Etonian), chairman of the newly formed Wessex committee, who had become a viscount when, after a spell as Secretary of State for Defence at the time of Suez had exhausted his taste for politics, he withdrew to become High Commissioner in Nigeria. In one of his vivid word pictures Norman described him as a 'renaissance man with strong opinions vigorously expressed. He did not like any Q.A.G.

247

interference and foretold doom to the whole system of dual control unless he got his way.'[12] He also had a fine throwaway wit and a left-wing wife. These were influential figures who advanced on the London office like Shakespeare's feudal barons, their authority reinforced by the Benson Report. Gascoigne has described his strategy:

> With my professional team we would often come up with excellent ideas for this region, to which I would then find entrenched opposition at headquarters in London from those professionals who felt their power was draining away. After discussion I would very much regret their disagreement with our proposals but would then refer them to paragraph 80 [of the Benson Report]. This announced that regional committees are autonomous except for: (a) acceptance of properties; (b) major financial decisions. So we were able to put our schemes into practice.[13]

By 1976, when committees had been appointed for all fourteen regions, their chairmen were drawn from a somewhat wider background. They included the first woman, Sylvia Gray, a former chairman of the National Federation of Women's Institutes, Michael Cadbury from the Birmingham chocolate family, two former ambassadors – Sir Evelyn Shuckburgh and Sir Fred Warner – and Sir John Winnifrith.

The whole of 1969 was taken up with discussing the Benson Report. The Council awoke from its apathy and met five times between 14 February and 25 July, eventually adopting Benson's principal recommendations; members' meetings were held up and down the country; and finally the AGM in November approved the Council's conclusions by 9749 votes to 23, including for the first time proxy votes.

The major changes had then to be incorporated in a new Act of Parliament. At this stage Robert Mowat, the chartered accountant who had supported Rawnsley, renewed his efforts to improve the presentation of the accounts and petitioned against the Bill. He was successful in gaining the support of David English, the Labour chairman of the parliamentary Opposed Bills Committee, and also of Captain Ralph Bonner-Pink, a Conservative Member. The Trust was forced to rewrite the accounting provisions and also to concede the right for extraordinary general meetings, another matter on which the reform group was not satisfied. In general, however, the Benson Report had gone a long way to meet their case and little was heard of them again.

The Benson Report proved to be a seminal document. Over twenty years later when Lord Oliver of Aylmerton, a former Law Lord, was asked to review the workings of the Trust's constitution he concluded that the 1907 constitution, as amended in 1971, had worked in practice. In discussing the rights of members he commented that it was doubtful whether 'a newly conceived constitution for a charitable trust in 1990 would provide for any public participation in its governance'. But the Trust 'is a unique voluntary organization constituted for the benefit of the nation and in which the subscribing members have, from the outset, been essential partners. No alterations to the constitution could today be contemplated which sacrificed that essential partnership element.'[14]

The task of examining the Benson Report's 600 paragraphs and recommending whether the conclusions should be adopted, amended or rejected fell to Sir John Winnifrith, who in January 1968 had been appointed the Trust's first director-general. The post of secretary was separated from that of chief executive, who was designated director-general, as Harold Nicolson had suggested more than twenty years earlier. Antrim and his colleagues felt that the job of chief executive had become too much for Rathbone. Soon after the extraordinary general meeting, at which Antrim had found him of little help, Rathbone suffered another nervous collapse. He himself recognized that a different sort of man was needed and he fell in with Crawford's suggestion that he announce that his early retirement was his own decision. This was generally believed and went some way to alleviating the disillusionment felt by some of the staff. But Rathbone never recovered from the blow. 'I mind going very much,' he said in a letter to Rosse. 'But I am sure that it is the end of an epoch of great expansion, that this is in the best interests of the Trust (this is all that matters) . . . . and the knowledge that Sir John Winnifrith, a splendid man, has been found does much to lessen my sorrow.'[15]

The decision to recruit a top civil servant was a sign that the Trust was indeed entering a new epoch. Winnifrith arrived after eight years as a powerful permanent secretary to the Ministry of Agriculture. Aged fifty-nine, he had followed the well-trodden path from Westminster School to Christ Church, Oxford, and subsequently enjoyed the classic career of the ablest civil servants, in the Board of Trade, the Treasury and the wartime Cabinet Office. In appearance he was suitably square and shaggy for Agriculture and 'could have come out of *Barchester Towers* – if not a bishop (his father had been a clergyman) then certainly a squire'.[16] The prospect of a bureaucrat, perhaps without much understanding of the Trust's aims, was at first viewed with some suspicion, but he soon won everyone's respect and affection. Despite all the changes introduced during his time he left behind no enemies, only friends. At the end of his three-year stint Antrim recognized that the Trust had been fortunate to have the service of 'a brilliant hard-working man whose training had increased his natural unselfishness [and] whose sense of the ridiculous had entertained all who worked with him'.[17]

Winnifrith had insisted that he would not stay for more than three years and in January 1971, at Gibson's suggestion, another eminent ex-civil servant was appointed as his successor – Frederick Bishop. Bishop was not a silver-spoon man. He had left a local grammar school in Bristol to enter the Inland Revenue at the age of nineteen, from where he took an external law degree. He was principal private secretary to two prime ministers (Eden and Macmillan) and then deputy secretary of the Cabinet. In 1964 he was made permanent secretary of a dud department, the newly formed and short-lived Ministry of Land and Natural Resources. As a result he left the Civil Service early to become a director of S. Pearson & Son. He was not entirely happy there but the experience at least had the advantage of making him known to Gibson, at whose suggestion he came to the Trust. Lithe and good-looking, he was much liked, and he interfered less than his predecessor. John Gaze, the author of a history of the Trust and chief agent from 1976 to 1982, described his unobtrusive style: 'That things went smoothly and with increasing rapidity was due in large part to the conditions he created. Provided that his colleagues were going in the right direction he did not interfere with detail, ensured that they were not unduly hampered by committees and gave them the backing they needed to get things done.'[18] The staff had reason to be grateful to Bishop for obtaining decent though by no means generous salary levels and for setting up reasonable pension arrangements.

There was one particular gain to the Trust for which Bishop was partially responsible. Despite the tax reliefs granted from time to time, tax was still payable on certain bequests and gifts. Bishop made 'repeated representations' to the Chancellor of the Exchequer, Anthony Barber, and the Prime Minister, Edward Heath, both of whom he knew well, for better tax treatment. The representations were successful. The Finance Act 1972 made all bequests to the Trust and other charities exempt from estate duty and relieved all gifts from capital gains tax. Had the changes come a few years earlier Aynho House in Oxfordshire would have come to the Trust and some gifts of cash would not have been taxed at 80 per cent.

Bishop needed all his diplomacy to implement the plans for decentralizing responsibility to the regions. The underlying objective was undoubtedly right if the Trust was to manage its growing estate. But the chain of command was confused by making the senior regional staff responsible to the regional chairmen rather than to the director-general by whom they were appointed. And there were occasional good-humoured battles between head office and the regions, particularly where the regional chairmen had been military commanders or government ministers not accustomed to being gainsaid.

One question not discussed by the Benson Committee was the division of responsibility between committees and staff. Despite the abolition of some head-office committees, Benson's call for better reporting on existing and prospective properties, coupled with the new regional structure, required much more paperwork and could result in delays. In February 1972 Acland wrote to Bishop apologizing for rearranging papers for the head-office committees. But he added:

> At least some of the difficulties which happen in my office as well as yours are of the Trust's own making. Let's face it, these three [minor] acquisitions have been considered by: 1. Staff in the Lake District, 2. Committee in the Lake District, 3. Staff in London, 4. Properties Committee, London, 5. Finance Committee, London, 6. Executive Committee, and the six considerations have come to the same conclusion on every occasion.[19]

From his experience in the Colonial Service, Jack Boles, who had been secretary since 1968, replied that a federal system of government 'leads to a good deal of duplication and wasted effort, but it nevertheless sometimes seems to be the best answer; we must learn to live with a system that at times is cumbersome'. It was evident that too many small decisions were still being taken by head-office committees and not enough authority had yet been delegated to staff.

The 1970s was a decade when Labour and Conservative governments alternated, with policies which differed more in detail than in concept. Both parties had to deal with problems of inflation and industrial unrest, symptoms of more deep-seated economic difficulties. Britain's entry into the European Community on 1 January 1973, confirmed by the referendum of June 1975, marked the final stage in moving from the position of an imperial power to that of a partner with continental Europe. And in February 1975 Margaret Thatcher's defeat of Edward Heath in the election for the Conservative leadership heralded the most fundamental change in direction since 1945, from Keynesian to Friedmanite economics and from consensual to confrontational politics.

The Trust's most pressing concern was finance. Income had not kept pace with responsibilities. Between 1959 and 1969 the number of historic houses and gardens open to the public increased from 130 to 200, the acreage from 240,000 to 350,000, membership from 80,000 to 177,000 and visitors from 1 million to 2.7 million. Total expenditure rose from £1.1 million to £3 million, but annual revenue, including legacies, rose only from £1.4 million to £2.8 million, transforming a surplus of £300,000 into a deficit of £200,000. In the 1970s expenditure was further increased by the administrative cost of implementing the Benson Report, estimated at £212,000 a year. In addition £4.75 million capital works were required for repairing historic buildings (£500,000 for Knole alone), improving farms and cottages, and providing amenities for visitors.

Continuing high inflation from the early 1970s to the early 1980s, exacerbated by the oil crisis of 1973, created further problems. Inflation rose to 24.2 per cent in 1975 and, after falling somewhat, then rose again to 21.9 per cent in 1980. The Trust's income was hard hit because of its dependence on rents and the effect of dividend control on investment income, while its expenditure rose rapidly because of the disproportionate increase in building costs and wages. Between 1973 and 1975 wages at properties went up from £984,000 to £1,582,000 and the deficit at properties went up from £434,000 to £753,000. The annual report for 1975 sounded a note of alarm: the Trust's 'future development, and possibly its very survival, appear . . . to depend no longer upon its own initiative, but upon the government's ability to bring inflation under control. No matter how severely expenditure is pruned, no matter how diligently new sources of income are exploited, a high level of inflation will make it impossible to maintain the standard of conservation which both members and the general public have come to expect from the Trust.'[20]

The man responsible for steering the Trust through these dangerous shoals was Mark Norman. With three director-generals and three chief finance officers during the decade, it was fortunate that he had taken over as chairman of the Finance Committee from his cousin, Holland-Martin, who had become less effective since being injured in a riding accident some years earlier and did not seem to have realized that it was no longer enough to manage the Trust's finances like the corner shop – 'Let's look and see what's in the till.' Norman's first task was to replace the chief accountant by R. J. Wheeler, not a qualified accountant but a man with administrative experience at Unilever, tactful, cheerful and numerate. Goaded on by Norman, he imposed a system of financial control on the regions, which were always in danger of over-spending. The new budgetary regime was achieved without losing the goodwill of the spenders – the agents and historic-buildings representatives – but in late 1974 their loyalty was severely tested.

In February 1975 the prospects looked so unpromising that the Executive Committee approved contingency plans for a cut in real expenditure of about a fifth for the following year. This could only have been achieved by closing parts of some houses and gardens, postponing repairs and improvements, and slowing down the rate of replanting woodlands. When Bishop outlined the proposals to a 'Doom and Gloom' meeting of senior staff, tempers ran high. Trinick and other optimists believed that if the Trust were to exploit its assets it could survive without making damaging cuts. In May 1976 (by which time Bishop had retired), Trinick gave a talk at the Agents' Conference entitled 'Is Our Financial Psychology Back to Front?', in which he outlined an alternative programme, aimed at living with inflation by keeping admission fees and rents in line with price increases, and raising income rather than cutting expenditure. In addition to urging the appointment of more trained recruiters, Trinick argued that the properties should make more effort to attract visitors and compared the Trust's record unfavourably with that of some private owners: 'The Properties Open book is a misnomer. Many of our most popular houses are closed. I don't think Lord Montagu or Lord Bath would be proud of an annual increase of 8 per cent [in visitor numbers]' (the Trust's figure for 1970-4).[21]

Despite continuing inflation, the more drastic cuts did not prove necessary – buoyant membership and visitor figures combined with postponing some building works and leaving empty some vacancies enabled the Trust to break even until 1977, when a record number of gifts and bequests enabled spending plans to be resumed. Bishop decided to take early retirement in May 1975 and was succeeded by Jack Boles, who had been the Trust's secretary since 1968 and had filled Bishop's place during a prolonged period of illness.

Boles had had a very different career from his predecessors. Born in 1925 and educated at Winchester, he spent three years in the Rifle Brigade before entering the Colonial Service, where he was posted for some years to North Borneo. In 1964 he returned to England at the age of thirty-nine with four children and no job. Seeing an advertisement for the post of assistant secretary he applied and was chosen in the hope that his training would enable him to bring some order into Rathbone's happy disorder. He became secretary under Winnifrith when Rathbone retired. After two outside appointments there were good reasons for the next director-general to come from inside the Trust, both in order to boost staff morale and to have a chief executive who knew the people and problems. Bishop strongly advised Antrim to pick Boles as his successor. Antrim quickly agreed, telling Rosse that 'his virtues of hard work and reliability are terribly important. . . . I believe he will do the job very well.'[22] Despite opposition from several grandees Antrim was proved right. Boles' firm leadership (at one time he was known as the 'sergeant-major'), high standards and friendly manner enabled him to consolidate the changes following from Benson.

Boles was quick to react to new opportunities. In the mid-1970s he took advantage of the Government Employment Schemes under which the Trust provided work and training while wages were paid by the Manpower Services Commission. These enabled the Trust to catch up with a vast backlog of repairs on the properties – repairing park walls, clearing scrub, planting trees – and also to recruit promising employees when regular vacancies occurred. At the height of the schemes in 1986 there were 3800 young people on projects costing a total of £11 million, to which the Trust contributed £1.4 million.

Boles had established a strong hold on the organization by 1977, when Antrim died while still in office. Antrim was followed by Patrick Gibson, who had been in line for the succession since Crawford's day. Like his three predecessors he was an Etonian and an Oxonian. He was the first businessman to become chairman of the Trust. After an eventful war (he escaped from an Italian prisoner-of-war camp) he joined S. Pearson & Son, becoming chairman of the newspaper and publishing side of the company in 1967 and chairman of the group from 1979 to 1983.

Gibson knew the Trust well, having been an active member of the Benson Committee, and of the Council and the Executive Committee until he resigned on becoming chairman of the Arts Council in 1972. This was a challenge he felt he could not refuse, though he told Rosse, 'The NT is what I really care about, not all those ghastly avant-garde plays. And when my five-year stint is up I shall be only too happy to return to the NT fold if they want me.'[23] By this time Antrim had died and they did. But Gibson wanted to be certain. 'I should love to take it on,' he wrote to Rosse,

> if the Council are sure I am the kind of man they want. . . . If they feel that they ought to have someone who seeks the headlines and 'makes news' then clearly I am not the man. There are those who feel the Arts Council needs that sort of leadership and that I have not provided it. Certainly I have not and I do not think the National Trust needs it either . . . though of course one must speak up when the occasion demands.[24]

Gibson's talents and experience were tailor-made for the Trust: he was a keen gardener, he had a wide knowledge of the visual arts, and on his regular visits to properties there was little that escaped his eye for detail. A neat man with an incisive mind, he handled meetings with finesse and good humour and he took trouble in getting to know members of his Council, who appreciated the lunch parties he gave for them at home. He had been made a peer in 1975 and so had a forum in which to represent the concerns of the Trust. Most important at this juncture in the Trust's history, he had all the skills needed to run a large organization.

CHAPTER TWELVE

The New Professionals

'AMATEURISM, IN the real and best sense of the word, seems to be the essence of the National Trust' was Bishop's description of the organization as he found it in 1971. Ten years later, almost every aspect of the Trust had become thoroughly professional. The increase in staff, from 933 full time in 1968 to 1488 in 1981, had allowed for expert advisers on nature conservation, architecture, archaeology and a host of other subjects from catering to paper conservation. The fact that there was money to pay these people was due to the increase in membership from 160,000 in 1968 to over a million in 1981, in itself the result of the new professional approach to recruitment.

The economic and intellectual climate had been highly favourable to the Trust since the early 1960s: the growth in national income, the reduction in working hours and the increasing ease of transport were bringing more visitors to Trust properties, and the new concern for conservation, both of historic buildings and of the natural environment, was inspiring more people to join environmental groups of all kinds. But there was no systematic campaign to recruit new members until the arrival in 1969 of Edward Fawcett, the Trust's first director of public relations. Fawcett selected himself. Since the age of seven, when he had seen a photograph of a property in the Lake District, he had wanted to work for the Trust. After war service in the navy, he had spent a year at the London School of Economics before joining Shell and then moving to Joseph Lucas Ltd, acting first as exhibitions and publicity manager, then as a director. During this time he had taken an external degree in the history of art, studying under Sir Nikolaus Pevsner and Sir John Summerson, in the hope of obtaining a post as a historic-buildings representative. But by the time he had arranged to see Fedden, whom he knew slightly, he had come to the conclusion that he could be more useful in public relations. He had read the Benson Committee's analysis of the Trust's failings and he had discovered for himself that it was impossible to become a member at Stourhead, the Trust's most visited property (there were no forms). Fawcett recalls his final selection being more like a sherry party than a job interview. He was summoned to Antrim's Chelsea home, where he also found Lords Gibson and Rosse, together with Fedden, Winnifrith and several others. After a volley of questions, somebody pointed out that there was a committee meeting to attend and everyone except Fawcett left for Queen Anne's Gate. A few days later he was appointed.

The Benson Committee had been critical of the Trust's public relations. The trouble was that the leaders of the Trust regarded the concept of 'PR' with disdain and were not prepared to allocate adequate resources to recruiting new members, making the Trust known to the general public, attracting more visitors to the properties and explaining its activities to members. Not only did Fawcett make an impact on each of these matters, he also handled the press well – his courtesy, in Norman's words, concealing 'a brilliant capacity to seduce the media'.[1]

When he arrived to take up his new post he was given no advice or instructions, though Antrim mentioned that he 'might like to look at membership'. Fedden on the other hand told him, 'You must realize that the Trust has nothing to do with people.'[2] Fedden believed that all the Trust needed was a couple of millionaires rather than a mass membership, though he did concede that the Trust might aim at half a million by the end of the century.

Fawcett set his sights much higher. Once he had arranged for application forms to be available at the properties there was an immediate response, but the forms could not be processed quickly enough. The membership department was based at the Blewcoat School, a small eighteenth-century building near the Trust's headquarters, where thirty or forty diligent women bashed away on their manual typewriters recording particulars on individual cards. It was a highly personal service: one entry noted that the member must be telephoned when her card was sent out because her dog would intercept the post and chew up the contents. Boles solved the bottleneck with his usual efficiency and in 1970 transferred the operation to a computerized system at Ravensbourne Registration Service in Beckenham. No doubt the lady with the dog suffered, but there were other benefits.

The increase in members involved two distinct operations: first they had to be recruited, and then they had to be retained. Michael Beaumont, who had a background in advertising, realized that more positive marketing of the Trust to the visitors could produce more members. Stands were set up near the ticket offices, the wives of Trust employees were trained to operate them and visitors were offered a refund on the entry fee. The formula was successful and recruits soon reached record numbers.

The retention of members was not easy during the inflationary years of the 1970s when subscriptions had to be raised substantially in order to keep up with costs. Many people paid by banker's order and, when asked to sign a new order for a higher sum, a sizeable proportion did not do so. The difficulty was overcome in 1977 when direct debit was introduced, making it possible for members to be notified of rises in subscriptions instead of having to sign a new form. This meant that subscriptions could be raised by smaller amounts more frequently, and it also resulted in lower administrative costs. In the same year, following extensive market research, alternative forms of membership were introduced for families, pensioners and under-twenty-three-year-olds.

Membership numbers took off as a result of these innovations. In 1970 the increase was 50,000 and for the next twenty years it seldom dropped below that figure, reaching a staggering rise of over 170,000 in 1978 and again in 1989 and 1990. Starting from 226,000 in 1970, the total topped a million in 1981 and two million in 1990. This made the Trust by far the largest of any charitable organization in Britain and one of the largest environmental organizations in the world.

It was the regional information officers who had to carry out these new plans. At the time of the Benson Report four were already in post – in Wales, Northern Ireland, the Lake District and the south-west. The attitude of the regional agents was crucial: in Devon and Cornwall Warren Davis, who in 1981 became the Trust's national press officer, found that Trinick was as keen as he was to open new shops, organize plays and concerts and get more publicity for the Trust. Having been a journalist and still a part-time newsreader at BBC Plymouth, Warren Davis was in a good position to do so. In the Lake District, Acland was more cautious: he allowed his information officer, Christopher Hanson-Smith, an ex-Colonial administrator, to set up information centres for recruiting members and selling maps and guidebooks and to organize occasional open days for members, but although he was himself a considerable showman he believed that his mission was to protect the landscape and keep alive the torch handed on by Beatrix Potter rather than to market the Trust or its properties.

Within two years of Fawcett's arrival, information officers had been appointed in most regions. One of their principal tasks was to capitalize on the growing fashion for visiting country houses and gardens and to ensure that the properties were open for longer periods, as the Benson Committee had insisted was necessary. An entirely new initiative was a programme of events, beginning soberly with Shakespeare plays and classical music, and gradually becoming more ambitious. The first of many *fêtes champêtres*, illuminated by floodlights and fireworks, with the guests encouraged to arrive in fancy dress, was organized at Stourhead in 1978 by an ex-actor, Alastair Bannerman, who was the animator of many subsequent occasions. There were also horse trials and veteran car rallies as well as a host of occasions for the public to meet the staff and be shown what was being done in the gardens and for nature conservation. Advertising, signposting and the provision of facilities at the properties all contributed to the large increase in the number of visitors from about 3 million in 1970 to 6.6 million in 1980 and more than 10 million in 1990.

Fawcett turned his attention next to trading, in the belief that many of the visitors would like to buy a souvenir and sip a cup of tea. The Trust had opened its first shop in 1965, and the first to be purpose-built in 1968, but most houses had merely a table in the front hall and the average sum spent was not more than two old pence. There was only a small range of goods for sale (guidebooks, postcards, badges, ties and a few other items), as the Committee of Taste which vetted new products had not approved a single object in five years. Fawcett found that even moving the goods from under the table to display them on top had a dramatic effect: it was clear that visitors were avid to buy something. Once a range of useable, well-designed and moderately priced products had been developed by Ray Hallett, the marketing adviser, and Pat Albeck, who had taken the pattern for the Trust's first tea towel from a carpet at Saltram, the new merchandise sold fast. The Trust had hit on the right combination of necessities and souvenirs – jams, toiletries, scarves and china – dressed up in designs of flowers or objects from the historic houses. At the time this was a new concept, though the idea has since been so widely copied as to become commonplace.

The shops were set up with the minimum of capital. The

sales staff were volunteers, the premises were former cloak-rooms, lobbies or stables. The goods were displayed on tables, dressers and clothes horses. Even tills were thought to be too expensive. The result was a new style of merchandizing, which the visitors loved, and the return on investment was more than satisfactory.

The first major difficulty arose when it was decided that the operation had expanded to a point which made full-time professional management essential, and the first director of trading, a retired colonel, was not a success. The Trust started to lose money, but Boles was reluctant to sacrifice him. In desperation Norman persuaded a friend with experience of retailing to supervise the shops, but within a few weeks he reported that either the Colonel must go the next day or he would. With the appointment of Roy Preece, an experienced trading manager, the business became more efficient, and was developed by Martin Moss, who had been managing director of Simpsons of Piccadilly. He introduced clothes and other goods to appeal to younger customers, adopted computerized reporting systems and even exported some goods across the Atlantic. By 1980 there were 135 shops, all but 12 at the properties, with a total trading turnover of £6.7 million, making a net contribution of £346,000. Ten years later the trading turnover had risen to £19.5 million, yielding a contribution of £5.1 million.

The Trust relaunched its book publishing in 1973 with the *National Trust Guide* edited by Robin Fedden and Rosemary Joekes. The first five books, including the *What to See* atlas and Graham Thomas' *Gardens of the National Trust*, each sold more than 120,000 copies, helped by big book-club sales, but later books published in association with other companies did not make the returns hoped for. In 1985 Margaret Willes, who had been editorial director of Sidgwick & Jackson, was chosen to direct the Trust's publications, one of the first women to be appointed to a senior management post. She decided that the

Trust must be master of its own fortunes and must establish its own imprint. In 1987 the *Annual Handbook* was published as an independent venture and the following year came the first major book, a new edition of the *National Trust Guide*. This was followed by a series of large illustrated books, cookery books and books for children. Margaret Willes' terms of reference were to raise the profile of the Trust, to inform the public and, if possible, to make a profit. In fact, at least up to 1990, her operation proved profitable, the new style of guidebooks, which combined a guide to the house with a social history and colour illustrations, being particularly successful.

The information officers quickly realized that decent lavatories must be provided and that good catering was necessary if visitors were to enjoy an afternoon at the Trust's houses. Elaborate menus were not required but the tea rooms had to be welcoming and had to retain their country-house atmosphere. Moreover the food, whenever possible, had to be home-made. The difficulty was to combine these standards with making a profit. In this as in so much else, Devon and Cornwall led the way. Sylvia Gray, chairman of the Trust's Regional Committee for the South Midlands, who ran the Bay Tree Hotel in Burford, was brought in to advise. The aim was to serve straightforward home-made food – soups, salads, cakes and pies. By 1980 there were about seventy tea-shops and restaurants and four inns.

In 1984, when retailing and catering had become sizeable businesses, a company was set up to operate the trading activities and covenant its profits to the Trust. Other enterprises included 175 holiday cottages which, if not needed for local housing, were let furnished. In 1990 the trading operations together made a profit of £5.1 million, amounting to 7.2 per cent of total income.

Paradoxically, as the staff became increasingly professional, so the Trust came to depend more on volunteers. From the early years the Trust had relied on the voluntary work of members of local committees to administer many of the open spaces, including Wicken Fen and Hindhead Common, but until the end of the Second World War no non-members were involved. Young volunteers who were not members first came to help in 1947, when a group from the Youth Hostels Association cleared away barbed wire left from wartime defences at Tintagel. In the early 1960s young volunteers came in large numbers, first to help with the Stratford-upon-Avon canal and then with Enterprise Neptune projects. The Trust responded, organizing camps, known as Acorn Camps, and providing simple accommodation, often in converted farm buildings. The volunteers accomplished a variety of projects, ranging from reroofing an old mill in Bransdale, Yorkshire and rebuilding stone walls in the Peak District to digging out ponds, clearing scrub and planting trees.

Older volunteers usually worked indoors on such different tasks as repairing textiles, lecturing and various professional and secretarial jobs. Most of the stewards in the houses were volunteers, often drawn from National Trust Centres or Associations, as these groups of keen members are now often described. The first Centre had been formed in Manchester in 1948 and was soon followed by others in London and Birmingham. The early ones were organized spontaneously and received little more than goodwill from head office, but later the regional information officers were able to help form new centres. At the time of the Benson Report there were nineteen centres and the number increased to 122 by 1980 and to 183 by 1990. As well as being a source of volunteers and of fund-raisers, the centres foster the interest of mainly older members through their programmes of lectures and tours, and facilitate closer relations with the Trust's staff.

The impact on the Trust's annual income of the larger membership was striking. Subscriptions, which rose from £400,000 in 1970 to £4.3 million in 1980 and £28.7 million in 1990, were made to produce more income for the Trust by persuading members to sign deeds of covenant, thus enabling income tax already paid to be reclaimed from the Inland Revenue. In addition legacies went up from £327,000 in 1970 to £6.9 million in 1979 and to £21 million in 1990. These increases compensated for the relative decline in rents and admission charges during the 1980s.

Fund-raising had become more expert and enabled the Trust to undertake many projects which would not otherwise have been possible. A lot of land was bought in this way in the Peak District, for example, and the 1987 Storm Disaster Appeal raised more than £2 million within a year. For the first time serious efforts were made to attract business sponsorship for a variety of projects, including the *Annual Handbook* as well as concerts and other events. Many individual benefactors, small as well as large, took a great interest in how their money was to be spent, calling in at the office to discuss the possibilities. Among them were Mr and Mrs Kenneth Levy, whose bequest of £5.6 million was the largest sum ever to be received. During their lifetime they had supported the restoration of Clandon Park and the Bath Assembly Rooms as well as the establishment of the textile-conservation workshop at Hughenden, and it was decided to set aside £3 million for similar purposes.

If Fedden and other members of the staff felt that the Trust had nothing to do with people, still less did they feel that it had anything to do with children. They argued that it was not the Trust's business to spend money on teaching and so ignored the opportunity to interest young people in the Trust's work and in conservation generally. Faced with opposition to the idea that children should actually be encouraged to visit Trust properties, Fawcett set out to prove wrong those who feared that children would harm valuable furnishings and objects. All the houses were asked to report any damage or nuisance: out of 75,000 visiting children, the worst culprit was one who had run his toy car across a marble table.

The argument was won and in 1979 the Trust appointed its

first full-time education adviser, John Hodgson. Hodgson had been deputy director of the Geffrye Museum in East London for some years, where he had seen how children could be inspired by aesthetic education and by visits to the country-side. At that time he wrote to Fedden suggesting that the Trust should emulate these pioneering ventures but received a courteous brush-off. However, in 1970 he was appointed to create a Museum of Childhood at Sudbury Hall in Derby-shire, funded by the county council. Here he had the oppor-tunity of 'insinuating education into the house', which he did without seeking permission from the Trust. Sudbury became the first historic building in the world to introduce a live educational programme as distinct from lectures.

In 1976 he discovered a local 'Theatre-in-Education' com-pany with whom he conspired to run an experiment, again without telling the Trust. It was a great success. Aided and abetted by Fawcett, he arranged for a production to be given at Lacock and invited a party from Queen Anne's Gate. They were bowled over – not by the educational potential but by the public relations opportunities. This was the beginning of the Young National Trust Theatre, which each summer since 1978 has produced a play on a particular theme and period for professional actors to join with children from local schools and perform at selected houses.

At first there was resistance to education in some of the regions. But in others the idea met with enthusiasm, and some of the house administrators and wardens were extreme-ly keen. With the help of information officers, a few local education authorities and a host of volunteers, school visits were arranged and literature provided for teachers and child-ren. When Hodgson retired in 1989 he had persuaded the Trust that children were important to its future and had passed on something of his own vision: 'I see the Trust's woods, parks and gardens, its wonderful buildings and their contents as the life-preserving Tree which can nurture the coming generation and inspire them to create a new Renaissance.'[3]

The Trust's new commitment to education was demonstrated by the appointment of the first education manager, Patricia Lankester, and her location at head office rather than in isolation at Lacock, where John Hodgson had been. The introduction of the national curriculum enabled her to tailor the Trust's facilities to meet the needs of each age group and each subject. Key properties were identified which could offer a variety of opportunities – Cragside, Fountains Abbey and Box Hill among them – and education officers were appointed in most regions. Within four years the number of school visits had doubled and by 1993 had reached 500,000.

CHAPTER THIRTEEN

Decorating Historic Houses

MANAGEMENT OF the Trust's historic houses showed as marked a move towards professionalism at the time of the Benson Report as did its public relations. The presentation, as well as the continuing care of the 156 historic buildings open to the public, imposed a great strain on what was then still a tiny staff; outside London there were only five full-time and five honorary historic-buildings representatives. Not until 1977, when Gibson became chairman, was the decision taken, on his insistence, that there should be a full-time appointment in each region. But in the absence of university courses in the decorative arts to match those in art history it was not always easy to find the right person.

Since 1956 Fedden had had the assistance of a highly expert adviser on pictures, St John (Bobby) Gore. After the army Gore had studied for two years at the Courtauld Institute under Anthony Blunt, who had been honorary picture adviser to the Trust since 1948 as to other still more august institutions. Against Blunt's advice Gore had spent the next five years at Sotheby's, and then, at Blunt's suggestion, he had been appointed to the Trust. For several years he was also representative in the Southern Region and in 1973 he succeeded Fedden as historic-buildings secretary.

Gore was less of an organization man than Fedden and played less part in the central direction of the Trust, but he was more of an aesthete and more in touch with the world of galleries. A gentle, smiling man, he told Fawcett that he hoped to bring curatorship up to museum standards, while maintaining the atmosphere of a private house. He was well aware that the V & A, where the furniture, textiles and interior-decoration departments were being revived, was critical of the Trust's standards. Despite his unworldly appearance, Gore was not without guile: he decided to disarm the critics by inviting such eminent scholars as John Pope-Hennessy, director of the British Museum, and Professor Michael Jaffé, director of the Fitzwilliam Museum in Cambridge, to join an Arts Panel set up to advise on the Trust's collections.

Gore's great contribution, taken much further by his successor Martin Drury, was to build up a professional conservation staff who could combine traditional historic-house practices with modern techniques. One of the by-products of their practical experience was *The Manual of Housekeeping*, written by Sheila Stainton, the Trust's housekeeper, and Hermione Sandwith, the deputy surveyor of conservation. The book was originally intended only for the guidance of staff but proved such a mine of useful information that in 1984 it was put on sale for the ordinary housewife, who must have been pleased to find that she was advised never to polish brass and furniture more than once a year. As time went on traditional care and repair was supplemented by electronic devices to monitor and control levels of light and humidity. For the public the more visible change was the lowering of blinds, essential to preserve textiles, watercolours and furniture for future generations, but sometimes rather deadening in their visual effect.

Another of Gore's innovations was to open a workroom for repairing textiles at Knole in 1974 where dozens of volunteers working under professional supervision restored the King's Bed, though even with their help the project took thirteen years to complete and cost £161,762. Later Martin Drury set up professional workshops at Blickling and Hughenden for the repair of historic textiles and the training of young conservators. The advantages gained in skill and economy from the Trust having its own facilities led to a workshop for repairing sculpture to be set up at Cliveden in 1983.

The Trust continued to use outside architects to supervise the structural repair of historic houses, but the restoration of the interiors presented problems. Until the 1960s decoration was usually left to the tenants, but then three run-down, rather empty houses were acquired – Shugborough in Staffordshire, Clandon Park in Surrey and Sudbury Hall in Derbyshire – and Fedden realized that they could not be opened just as they were. John Fowler, the leading decorator of his day (he founded the firm of Colefax and Fowler with Lady Colefax) and in some eyes the greatest since Robert Adam two centuries earlier, had already been involved in Claydon. He was coming to the end of his career and was prepared to accept a modest fee for advising on how to make these 'empty barns' attractive for the public. Slow of speech and schoolmasterly in manner, he inspired awe in his assistants. In his private work he was at his best with sparky women and at the Trust he got on well with the firm and knowledgeable St John Gore, then the representative in charge of Clandon, where he had the opportunity to plan a full restoration.

The house was designed in 1720 for the second Lord Onslow. The sixth earl lived there for a few years after the Second World War but soon gave up the struggle. After the house had lain empty for a few years it was given to the Trust by his aunt, the Countess of Iveagh. Only family portraits, some other pictures and a few large pieces of furniture remained, and Gore thought that the house could be given a new lease of life by introducing the porcelain and furniture collected by Mrs Hannah Gubbay and bequeathed to the Trust on condition that it was housed near London. The collection, rather small in scale and exotic in character for Clandon, was accompanied by a handsome endowment (most of which was swallowed up by taxation), intended to be used for its display.

Fowler proceeded carefully with the restoration, taking scrapes of the layers of paint underneath the white coating which had been applied between the wars. Wherever possible the white paint was removed to reveal the early colours; in the green drawing room the original wallpaper was exposed beneath the silk hangings and in the state bedroom a new paper was made to match fragments from behind the panelling. In other rooms Fowler based his colours on the scrapes, and where there was no clue to the original decoration he used his own judgment. Living only half an hour's journey from Clandon, he spent countless days in its rooms, evolving his distinctive blend of authenticity and innovation. The restoration was generally judged a great success, although the family thought it a disaster. Twenty-five years later it is one of the few Fowler interiors to have been preserved intact and has itself become an historic interior.

At Sudbury Hall, where Fowler was working at the same time, the problem was also to add interest and give a sense of unity to a house devoid of much of its contents. After taking the usual scrapes, he painted the woodwork of the staircase white, a decision that was criticized by outside experts, though he was vindicated later when further evidence confirmed his original judgment. Where there was no historical evidence, Fowler used colour to emphasize the exceptionally fine plasterwork and the architectural quality of the house – Italian pink in the great hall and a strong yellow on the walls of the staircase. Other rooms were left largely untouched, but Lord Vernon, who had handed over the house to the Trust, regretted:

> the deliberate change which has been wrought in the character and personality of the house. . . . What has happened is that a large amount of public money has been spent . . . to create an interior which they [the Trust] think, regardless of history, is aesthetically satisfying. But is this right and is it the true purpose of the Trust? There are large

numbers of the public . . . who prefer to see a house such as this as it had evolved through the centuries – warts and all.[1]

John Cornforth, who had become *Country Life*'s principal writer on historic houses and was a member of the Properties Committee, summed up Fowler's qualities: 'He was the only person in the country with the combination of historical under-standing, practical experience and ability to inspire, instruct and cajole all those with whom he came into contact. . . . his ability to open eyes and light fires of enthusiasm was just as important to the Trust as what he did, and his influence came at a crucial time.'[2] The Duchess of Devonshire described how he 'cajoled' his client:

He had a clever way of making sure the right stuff of the right colour was chosen for covering furniture and making cur-tains for a National Trust house where a committee was meant to do the choosing. Knowing that he was the one who knew, he was determined, rightly, to make decisions, and he bowed to the committee system like this: he produced several patterns and laid them out side by side. Picking up the first he said, 'I expect you think this too pale,' the second, 'I'm sure you think that blue is not right for the date,' and the third, which he had already chosen, he would grasp in his hand and say with a note of triumph in his voice, 'And *you*, darling, *you* with your *unerring eye*, will say this one is *perfect*.'[3]

After 1968 Fowler's health began to decline and the Trust occasionally made the mistake of asking him to do too much. After visiting Fenton House in Hampstead in 1974, Antrim risked one of his rare aesthetic judgments: 'It is not up to our standards. . . . The choice of colour for the walls and curtains in the dining room is in my view positively ugly. . . . We have used John Fowler for many years as the only person who knows how to decorate houses. He is now fairly old and extremely ill and I think that he has lost his touch and that the time has come to use someone else.'[4] In reply Gore acknowledged that Fowler's employment had led to over-indulgence in decoration, but he defended the work Fowler did when properly briefed. Gore then restated the Trust's guidelines: '(i) resistance to redecoration unless it is forced on us; (ii) repetition where possible of the existing scheme, with advice on obtaining colours, mixing paints, etc., from people of experience . . . . and (iii) in the case of an architectural interior consulting Fowler for the present' and in future using a member of a panel or committee if no suitable successor could be found.[5]

In fact a successor was found in David Mlinaric, who was engaged to prepare Beningbrough Hall, a splendid Baroque house in Yorkshire, for portraits of the period from the National Portrait Gallery. There were no documents to guide the restoration and the panelled rooms had been so thoroughly stripped in the 1920s as to leave little, if any, trace of the earlier paint. Mlinaric based his scheme on colours found elsewhere in the house and made his decisions on the spot, advised by an unofficial committee consisting of Gervase Jackson-Stops, John Cornforth and David Howarth, the Yorkshire representative. The subtle stone tones of the entrance hall, staircase and corridors, leading to the stronger tones of the principal rooms, were successful in bringing back to life a house which for some years had stood forlorn and almost empty, and in creating an appropriate setting for the 120 portraits on loan.

But Mlinaric could not devote enough time to the Trust and it was therefore necessary to devise a new approach. When in 1979 Tom Helme was engaged to advise on repainting and to carry out the more important projects himself the arrangements proved better suited to the guidelines which Gore had defined. Helme had exactly the qualities required, combining an academic training in art history with the practical skill of a craftsman. The concept of 'redecoration' gave way to a lower-key 'make do and mend' approach based on research into how the rooms had been painted and furnished at different periods of their history. This historical advice was the responsibility of Gervase Jackson-Stops, who in 1972 had been taken on as a research assistant and three years later was appointed architectural adviser, a post for which he was well qualified. He had studied history, including architectural history, at Oxford and had then held a three-year studentship at the V & A. His experience there was helpful in promoting greater harmony between the two institutions. The golden prose in which he presented the case for historic houses persuaded even the more hard-headed members of the Executive Committee.

Jackson-Stops' first major task was to work on the family papers at Erdigg, a late-seventeenth-century house in North Wales. Erdigg came to the Trust after seven years of tortuous negotiations with the donor, Philip Yorke, following his brother's death in 1966. He was a whimsical and eccentric man but he was determined to see that the house was preserved. There were two major problems: the lack of an adequate endowment and severe subsidence caused by workings from the coal-mine near by. Some people thought that the house was in such a state of dereliction that the Trust should not risk taking it on (the northern end had sunk by three feet, the southern end by five feet six inches and rain poured in through the middle), but Fowler was strongly in favour: 'this is just to add my fervent conviction that if Erdigg can be saved for the nation I know of no other comparable house and its contents surviving since the tragically sad disintegration of Uppark'.[6] He told Fedden that the atmosphere of a house had not moved him so much for forty years. The Coal Board did not stop mining until 1970 and the Trust found it nearly impossible to obtain precise information about the extent of the estate and its income. After endless difficulties, some of his own making, Philip Yorke became so exasperated by the delays that he threatened to leave everything to the Welsh Nationalists.

Eventually the Coal Board agreed to pay compensation for the subsidence and an endowment was found from the sale of an outlying part of the estate for development. But when the time came Philip Yorke could not bring himself to sign the deed of gift which the Trust's solicitor had brought from London. It was only on the third occasion, perhaps intimidated by a visit the previous evening to *A Clockwork Orange* (a film in which an elderly man is terrorized by youths) under the escort of Gervase Jackson-Stops and Merlin Waterson, the historic-buildings representative, both under the age of thirty, that he did so.

The restoration of the house took four years and marked two significant innovations. It was the first time that records covering more than 250 years were studied and used as a basis for the restoration and presentation of a property; and it was the first time that attention was focused on the daily life of the staff as well as the family. Unusually full evidence of the 'upstairs–downstairs' relationship had survived in letters, in verses composed by a succession of squires, in artefacts and in the buildings – the servants' hall was hung with eighteenth-century portraits of the estate and household staff and there was a complete range of outbuildings. Fascinated by the domestic history (described in his book *The Servants' Hall*)[7] Waterson, who was in charge of the complex restoration, decided that visitors should enter by the back yard, seeing the kitchens and laundries before the state rooms. This proved a popular and stimulating experience, but one that is rarely practicable and has not been repeated. With the remaking of the garden and the opening of a visitors' centre in the estate yard the property, which is within easy reach of Liverpool and Manchester, attracted thousands of visitors.

This broader approach, extending to the estate as well as the mansion, was taken further when Wimpole Hall in Cambridgeshire came to the Trust in 1976. After the property had been accepted it was discovered that the derelict home farm had been designed by Sir John Soane and as a listed historic building would have to be repaired and put to an appropriate use again. The regional director, Michael Rogers, suggested that it should become a centre for preserving rare breeds of farm animals. A benefactor came forward to give cattle, Suffolk Punch shire horses and funds for repairs and other construction work and for managing the animals. The great thatched barn was made into a museum of farming life on the estate and a special area was set aside for school parties. The home farm soon attracted more visitors than the mansion and they needed the usual facilities. Fortunately, the handsome but decaying nineteenth-century stable block had not been demolished, as some people had advocated. It was adapted for a shop and restaurant and a small part was returned to its original purpose – as stabling for the heavy Suffolk Punch carthorses, even if not for carriage horses.

A year later Cragside in Northumberland was accepted, more on account of its grounds than its house. The mansion had been designed for Sir William Armstrong, the inventor and armaments manufacturer, by Norman Shaw in a highly picturesque Tudor style. Some committee members thought it a Victorian monstrosity but Antrim was in favour and the question of late-nineteenth-century architecture had been settled in principle by the acceptance of Standen in Sussex several years earlier. (This was built for a London solicitor's family by Philip Webb with furnishings from the William Morris workshop and had been accepted on the advice of John Cornforth to the accompaniment of a campaign led by Sir John Betjeman and Sir Nikolaus Pevsner.) The most unusual feature of Cragside was its hydro-electric power system, with its dams and turbines, constructed by Armstrong. It was the first house in the world to be lit by electricity from water power and to have a number of other labour-saving machines, such as electric gongs to warn guests of approaching meal times. The Trust gradually restored many of

Armstrong's structures and constructed a path for visitors to walk round the 'Power Circuit'.

Cragside was transferred through the Land Fund, its running costs partly dependent on an annual subvention from the Historic Buildings Council, despite the Trust's resolve to accept no more houses on such unsatisfactory terms. However, it was the last of these arrangements; in 1980 the National Land Fund was replaced by the National Heritage Memorial Fund, which was given the power to provide endowments.

The change followed the Labour government's refusal in 1977 to use the Land Fund for the acceptance of Mentmore Towers, a mid-nineteenth-century mansion designed by Sir Joseph Paxton to display Baron Meyer de Rothschild's spectacular collection of paintings and other works of art. The outcome did not even save much money for the Exchequer: three paintings were bought for public galleries at a price of £1.4 million, whereas the property had first been offered to the government for £2 million and was eventually sold for a total of £9.4 million.

Mentmore was hardly typical, but in the 1970s the future of country houses in private ownership was once again under threat owing to the additional burden of capital gains tax and increases in other taxes combined with rocketing land values and repair costs. Their plight was examined by John Cornforth in a report for the British Tourist Authority, and in 1975 was portrayed in the V & A's exhibition, 'The Destruction of the English Country House'. The owners responded by founding their own trades union, the Historic Houses Association, to campaign for a tax regime which would allow private owners to maintain their houses in return for public access. With support from the Trust, the amenity societies and the Historic Buildings Council, they were successful in obtaining important changes to the Finance Acts of 1975–7: conditional exemption from the new capital transfer tax was granted for outstanding historic buildings, and maintenance funds could be set up for their support.

Following the public outcry over the Mentmore débâcle, the House of Commons Public Expenditure Committee decided to examine the working of the Land Fund. In giving evidence the Trust reiterated the view it had put to the Gowers Committee thirty years earlier – that owners should be encouraged to stay in residence. Where they were not able to carry on, the Trust was available as a safety net but would in future not be able to accept properties without an endowment unless they were of quite exceptional importance. The Expenditure Committee produced an influential report recommending that the Land Fund should be wound up and replaced by a National Heritage Fund vested in independent trustees who would have power to use the money as they saw fit for purposes within their terms of reference. The idea of the new fund was adopted in a White Paper published during the last days of the Callaghan government and legislation to set it up was enacted by the Thatcher government in 1980.

The National Heritage Memorial Fund (NHMF), as it was named, was established with wide terms of reference: 'to give financial assistance towards the cost of acquiring, maintaining or preserving land, buildings, works of art or other items of outstanding importance to the national heritage'. Equally important, it was free to make its own decisions and (by means of wise investment) it was not solely dependent on annual government grants. Public expenditure was incurred when payments were made into the Fund, rather than when the Fund paid out grants; hence it could act quickly and, if the resources allowed, could produce a large sum at short notice without having to seek government approval. Under its first chairman, Lord Charteris of Amisfield, Provost of Eton and former private secretary to the Queen, who seemed to have an almost instinctive understanding of what constituted the 'heritage', the NHMF interpreted 'the national heritage' with imagination: within its first two years grants were made not only for Canons Ashby, the Dryden family home, but also for the oldest-surviving submarine and for the greater horseshoe bat among other items.

The NHMF made a number of large grants in order to preserve historic houses with their contents intact, acknowledging that this was the principal purpose for which it had been created. Charteris himself took the lead in negotiating solutions for several houses which would otherwise have been sold up and in offering grants for objects of art to remain in houses from which they would otherwise have been removed. In its first year the Trust was helped to buy a painting by Bellotto which had hung in Powis Castle since the late eighteenth century but which had not been given with the property and was to be sold. The National Gallery's attempt to buy the painting was defeated, an outcome that was welcomed in the Principality, where it was felt that not everything should go to London.

The setting up of the NHMF coincided with the appointment of Martin Drury as historic-buildings secretary. After a spell as a Lloyd's broker and then eight years with Malletts, the dealers in fine furniture, in 1973 he became the representative in Kent and East Sussex, a post which he combined with that of furniture adviser. Drury has an almost visionary conception of the built heritage and its place in the wider landscape. It was he who first saw the need for a survey of the Trust's vernacular buildings and later for special care of endangered barns. In the restoration of buildings and their contents he insists on the highest standards, even at the cost of some delay, and his sure eye guides as much of the continuing work as he can take in on his whirlwind tours. His gift for understanding people of all sorts gives him more than anyone of his generation a feel for the pulse of the Trust.

During the 1980s four historic houses were transferred to the Trust with capital endowments which could not have been provided by the old Land Fund (except in Ulster). Because the Trust's properties are held inalienably and in perpetuity, the endowments allow for a capital sum twenty or thirty times the estimated deficit, an amount which can rarely be offered by a donor.

That was certainly true of Canons Ashby in Northampton-shire. The house was still owned by the Dryden family, by this time living in Zimbabwe, and although let it was gradually falling into dereliction – indeed one of the walls collapsed during repair. In May 1980 Jackson-Stops, who lived near by, heard that there was to be a sale of the remaining contents and, realizing that this was the last chance to save the place as an historic entity, he galvanized his colleagues into action. Embol-dened by the possibility of an endowment from the newly formed NHMF, the Trust most unusually decided to take the initiative. Jackson-Stops intercepted John Dryden on one of his rare visits to England in the hope of persuading him to make a gift of the house and the adjoining church. This he did. The Monument Trust offered the money for the contents and the NHMF offered £1.5 million for repairs and endowment, making possible the rescue of an extraordinarily romantic sixteenth-century house, first offered to the Trust over forty years earlier, and for which the Historic Buildings Council had long sought a solution.

Two years later the Trust engaged in campaigns to save two houses simultaneously – Belton House in Lincolnshire and Calke Abbey in Derbyshire. The circumstances were quite different: the trustees of the Harpur-Crewe estate had offered Calke Abbey to the Trust, accompanied by an endowment in the form of a large estate, in lieu of death duties, but their legal advice was at fault and the transaction languished. In the case of Belton, Lord Brownlow, who had moved to Paris, decided at short notice that, in order to meet his tax obligations, he had to sell what is probably the finest Restoration house in England and which retained paintings and furniture, many of them acquired for or long associated with the building. In private the family complained that the Historic Buildings Council had failed to provide help with repairs and that if the Trust 'could get together £2,500,000 for a tumble-down Canons Ashby' they ought to be more interested in a great house like Belton. Again it was the existence of the NHMF that enabled the Trust to approach Lord Brownlow and persuade him to suspend the sale while efforts were made to interest the Fund. Lord Brownlow offered the house to the Trust, and the Fund moved quickly to make available the £8 million needed for the endowment, the park and the contents, including furniture removed from the Speaker's house in the Palace of Westminster during the eighteenth century. Part of the village and surrounding land were acquired later in order that the core of the estate should remain together.

Calke Abbey was a more contentious case. It was not an outstanding building architecturally nor were its contents of outstanding artistic merit, but it had hardly been touched for a hundred years, it was set in a large park of exceptional beauty and, above all, nothing had ever been thrown away: it was a 'time capsule'. There was an early-eighteenth-century state bed with brilliant Chinese silk hangings which had never even been put up; a fine harpsichord had been abandoned in a stable loft; and any drawer might contain a 1924 cigarette card, an eighteenth-century pamphlet, a Victorian glove and a manuscript by Haydn. The house belonged to a reclusive family who had not admitted cars until 1924 and had made it one of the most secret places in England.

Charles Harpur-Crewe, who owned it for thirty-two years, died suddenly in 1981 while setting mole traps in the park. He had made no arrangements to protect the estate from death duties and his brother Henry, determined to keep it intact, offered it to the Treasury, intending that it should be transferred to the Trust. But for some reason the Treasury refused to accept the estate offered as an endowment and the Trust could not take it without one. After Save Britain's Heritage had led a spirited publicity campaign, Charteris took the initiative in

bringing the interested parties together and obtaining offers of
£1 million each from the Harpur-Crewe trustees, the Historic
Buildings Council and the Trust (helped by an anonymous
benefactor). With this on the table he managed to persuade the
Chancellor of the Exchequer, Nigel Lawson, to meet the
shortfall of £4.5 million and thus honour the government's
pledge that, whenever anything 'really big' loomed up, additional help could be sought. The physical work of rescue then
began: after four years of intensive repairs to the buildings, the
grounds and the contents, the house was opened to the public
in 1989 – though much was still left to be done. At Calke the
aim was to look as if nothing had happened: dirt was mixed
with the new paint, patching was preferred to repainting and
even cracks were left unfilled.

In 1983 the Trust was offered Kedleston in Derbyshire, the
finest neo-classical house in Europe, its saloon suggested by
the Pantheon in Rome, and still with its eighteenth-century
furniture – 'settees supported by gilt fishes and sea gods
absurdly like the king's coach' – and Flemish, Italian and
French paintings. Two years later, the more restrained furniture made by Thomas Chippendale for Nostell Priory, Yorkshire, which had been given to the Trust twenty years earlier,
was offered to the Treasury in lieu of tax on condition that it
remained in the house. The sums involved were far beyond
the resources of the Trust or the NHMF, which in the same
year was also asked to find a solution for Weston Park in
Staffordshire. The government again stepped in to help with
something 'really big' and made an unprecedented – and so far
unrepeated – grant of £25 million, which enabled the NHMF to
secure the future of all three 'entities' of house and collection.
Kedleston alone cost the taxpayer £13.5 million after the house,
park and £1 million had been given by Lord Scarsdale and the
Trust had promised to raise £2 million.

The acquisition of the only other country house to come to
the Trust in the remaining eight years of its first century,

Chastleton in Oxfordshire, was also made possible by the NHMF. In this case, as in that of some others, the transfer was as dependent on their diplomacy as on their funding, a total of £4.15 million.

CHAPTER FOURTEEN

Fora, Fauna and Flora

THE LAND agents provided the backbone of the Trust's management from the end of the Second World War until well after the Benson Report, when they were gradually supplemented, although not supplanted, by other professional staff. The size of the job they were expected to tackle is illustrated by Ivor Blomfield's experience when he joined the Trust as a young agent in 1959. He was sent to assist Colin Jones, the agent responsible for the whole area from Bristol to Manchester and worked from two small rooms above a shop in Ross-on-Wye, shared with a secretary and a bookkeeper. The only other management staff were two representatives, one in Cheshire and the other in Gloucestershire. Blomfield and Jones had to manage the agricultural estate, the commons and the woodlands and to look after the repairs, maintenance and staffing of the historic buildings and gardens – no mean assignment when so many new properties were being acquired and there were meetings of local management committees to attend in the evenings.

It was not until 1974 that the Trust's first botanical adviser was employed. This was partly because the agents were countrymen and were thought more likely to know about nature than about historic buildings or gardens, for which professional advisers were engaged much earlier.

Another reason was that, before the introduction of advanced technology, farming was generally assumed to be a sufficient protection for wildlife, except in the most important nature reserves. Until the late 1960s the wider effects of pesticides and fertilizers were not understood even by informed opinion, whereas the dramatic increases in output coming after a prolonged period of food shortages and rationing were naturally welcome. The average yield of wheat for example jumped from about a ton an acre in 1945 (only about a third more than in 1800) to almost three tons in 1982. The effects of new agricultural practices on the labour force and the landscape were equally dramatic and much more visible. Farm labourers were displaced by the new machines, and hedges were uprooted to allow room for the combine harvesters to manoeuvre; but it was some time before people realized how many birds, animals and plants had also been displaced.

The post-war Labour government's legislation for protecting the countryside had by no means entirely ignored nature conservation. But the aim was to protect specific nature reserves, rather than wildlife more widely, as John Dower's 1945 report on the National Parks had proposed. The National Parks Act of 1949 set out the framework for National Nature Reserves which would 'preserve and maintain as part of the nation's national heritage places which can be regarded as reservoirs for the main types of community, and kinds of wild plants and animals represented in this country'.[1] Within these reserves the prime objective would be to safeguard and, if possible, to enhance the nature-conservation interest. They were to be supplemented by Sites of Special Scientific Interest, which would be protected against agricultural and forestry development and would continue in existing land use. The work of designating sites went ahead steadily, but it was not until 1977 that the Nature Conservancy Council published a far-reaching report, *A Nature Conservation Review*, which identified the most important sites.

The Trust had been concerned with nature conservation since its early years. Its sixth property, acquired in 1899, was Wicken Fen, Cambridgeshire, the only remaining part of the

Fens which had not been drained and at which historic management practices continued. In 1907 the National Trust Act extended the original articles of association to include as one purpose of owning lands 'the preservation (so far as is practicable) of their natural aspects, features and plant life'.[2] A number of properties were bought by public subscription or were given because of their natural-history interest. They included part of the Cheddar Gorge, Somerset, in 1910, Blakeney Point, Norfolk, in 1912 and the Farne Islands off the coast of Northumberland in 1925.

In 1938 the Trust appointed an Advisory Committee on Natural History, which was chaired by Sir Edward Salisbury, a distinguished biologist and director of Kew Gardens. This lasted until 1957, when, in an attempt to sort out the Trust's organization, it was abolished because 'there was nothing for it to do'. In fact the Committee had met about three times a year to advise on such matters as the control of vermin, the spread of brambles and the reintroduction of particular species. In 1946, for example, it opposed reintroducing great bustards at Stonehenge, but a few years later it favoured reintroducing the large copper butterfly at Horsey in Norfolk. The Committee meetings were also occasions for reporting unusual sightings: in 1954 the chief agent had noticed cuckoos in pied and grey wagtail nests within a few feet of each other. The redoubtable Miriam Rothschild, the daughter of Charles Rothschild, one of the donors of Wicken Fen and herself a great naturalist – like her father, a world expert on fleas – had joined the Committee. As critical as her father had been of the Trust's attitude to nature conservation, she was shocked when the Committee was abolished, but the other members and the scientific organizations which made nominations to the Council seem to have made little or no protest.

It was unfortunate that the Natural History Committee was discontinued just at the time when the public was becoming more interested in environmental protection: the membership of environmental organizations roughly doubled in the 1960s and then redoubled in the 1970s, reaching about two and a half million in 1980. Among these organizations were the County Naturalists' Trusts, some of which, along with the Nature Conservancy Council, were becoming critical of the Trust's lack of interest in nature conservation, comparing it adversely with the National Trust for Scotland. In the early 1960s the latter had recruited graduates as naturalist rangers on the American pattern for some of its mountain properties; and on Ben Lawers it had opened a wildlife information centre. In 1969 the Scottish National Trust received a public tribute from Sir Fraser Darling in his BBC Reith Lecture for leading 'the world in the wholeness of its approach to environmental management'.[3]

In the same year the Trust appointed a new chief agent, Ivan Hills. He had some sympathy with the critics, believing that the committees had given insufficient weight to nature conservation. He was a great ornithologist and cared deeply for the countryside. He set about strengthening the natural-history side of the Trust's work, appointing J. H. Hemsley to be the Trust's first botanical adviser, reviving the Nature Con-servation Sub-committee and working more closely with the Countryside Commission. It was on his initiative that in 1976 the Trust opened a visitor centre on Witley Common to explain the wildlife of the Surrey heathlands.

The publication of the Nature Conservancy Council's *A Nature Conservation Review* marked a turning point in attitudes and provides a suitable moment at which to assess the Trust's record up to then. The *Review* showed that the Trust owned 200,000 acres of land of importance in nature-conservation terms, including 342 Sites of Special Scientific Interest. These were mostly in the uplands of the Lake District and Wales and on the coast but also covered a number of commons which the Trust had owned for a long time as well as the Farne Islands and other sites which had been managed as nature reserves.

Until the war, these reserves were, like other properties at the time, managed by local committees. After 1949 sites known to be of scientific interest were often leased to the Nature Conservancy or the County Naturalists' Trusts. Some of the pre-war local committees were highly effective, as was that at Scolt Head, Norfolk, where London University biologists were involved and seasonal wardens employed.

But even the best scientific advice was not a guarantee of good management if effective direction and adequate resources were lacking, as was the case at Wicken Fen. Towards the end of the nineteenth century peat digging and sedge cutting, which had shaped the Fen for centuries, gradually ceased. At the time the Fen came to the Trust 'preservation' was considered a sufficient means of protection. The Cambridge scientists responsible for the site shared this view until in 1929 Arthur Tansley, the eminent Cambridge biologist and later the Trust's honorary adviser on ecology, published a study showing that the ending of traditional methods was resulting in the growth of scrub. Without proper supervision of the local family who were employed as wardens or money for extra labour, the scrub gradually became a wood and the insect life diminished until in the 1960s visiting entomologists asked the Nature Conservancy to take over the site. The Trust was shamed into action and appointed as warden Lieutenant-Colonel Mitchell (whose son Peter became regional director of South Wales in 1985). He attacked the scrub with the help of volunteers, and a visitor centre was built to provide information for the numerous parties of students and schoolchildren. It was not until the early 1970s, however, when John Harvey, then a Cambridge lecturer in plant ecology (from 1986 chief nature conservation adviser to the Trust), began to use the Fen for teaching and research, that scientific measures for conserving the Fen's communities of plants and animals were introduced.

Another site not far from Cambridge which was not well

managed from its acquisition in 1924 until the late 1970s was Hatfield Forest in Essex. Its history has been studied by Oliver Rackham, the historian of the countryside, who describes Hatfield as:

> of supreme interest in that all the elements of a medieval Forest survive: deer, cattle, coppice woods, pollards, scrub, timber trees, grassland and fen, plus a seventeenth-century lodge and rabbit warren. As such it is almost certainly unique in England and possibly the world. . . . The forest owes little to the last 250 years. . . . Hatfield is the only place where one can step back into the Middle Ages to see, with only a small effort of the imagination, what a forest looked like in use.[4]

It is a forest not in the modern sense of planted trees but in the medieval sense of an area of land where the king or some other magnate had the right to keep and kill deer.

Hatfield was given to the Trust in 1924 by Edward North Buxton, a leader of the Commons Preservation Society and an early member of the Trust's Council. Right at the end of his life, hearing that it was for sale, he decided to buy it and give it to the Trust. The local committee, which included Buxton's sister and later his son Anthony, did not understand the nature of the forest nor how it should be managed. Almost immediately a number of ancient trees were felled; fencing was not repaired and cattle got into the coppices; rabbits and deer were decimated, regardless of the fact that they were one of the reasons for the medieval forest; and coppicing,* which had been carried out for centuries, became thinning, which was only appropriate for modern plantations. In Rackham's view this treatment was 'as if the Trust had been given an ancient dilapidated building and had maintained the corrugated-iron roof, mistaking this for a historic feature'.[5]

In 1955 the Trust's head office made a drastic decision: to

---

*Coppicing is a way of managing trees to produce regular crops of small-diameter wood by cutting just above the ground level every ten to twenty years.

lease part of the forest to the Forestry Commission for felling and replanting. Antony Buxton wrote to *The Times* and announced his resignation from the local committee in protest. In reply Rathbone claimed that the forest was lacking in young trees and had become impenetrable and that this could be remedied only by felling and replanting, which the Trust lacked the money to do itself. This scheme was shelved but four years later fifty acres of ancient coppice were grubbed out and replanted. Then in the early 1960s when the Trust was trying to improve access for the public, some of the woods were 'opened up' by bulldozing swathes through the coppices rather than by restoring the ancient rides.

Wicken Fen and Hatfield Forest illustrate the difficulty of managing sites where traditional husbandry has ceased and there is no historical study such as that published later by Rackham. At the end of his book he acknowledged that by the mid-1980s the Trust had mapped out an intelligent plan for the forest's future, and this has since been followed.

The 1977 *Nature Conservation Review* made the Trust realize that if it was to retain control of a significant proportion of its estate it would have to take a more direct interest in conservation. As a result, the Council decided that in future nature reserves would not be leased to other organizations except in special circumstances. By this time the Trust was in a position to manage them itself. Katherine Hearn, a botanist who worked for the Nature Conservancy Council, had been appointed to take charge of a biological survey of the Trust's estate. Moreover, by 1979 there were about a hundred wardens on the open-space properties able to carry out the measures recommended by the survey team – not yet enough, but a good start.

The biological survey was a mammoth task and took eight years to complete. Its findings were (at least in theory) incorporated into the management plans for each property and formed the basis for the Trust's work in protecting wildlife.

One of its most important and expensive recommendations was that traditional forms of management should be reintroduced in some of the places where it had lapsed, often before the Trust acquired the site. One of these was Box Hill, Surrey, where sheep grazing was restored and had a dramatic effect on the vegetation and wildflowers. Elsewhere 'creative conservation' was adopted, as at Kinder Scout in the Peak District, where sheep were excluded in order to encourage regeneration of heather and attempts were made to revegetate large areas of bare peat.

The growing concern about nature conservation was not confined to designated sites. In 1960 the Nature Conservancy had begun to investigate the effects of pesticides on wildlife generally, and two years later Rachel Carson's polemical best-selling book *Silent Spring* discussed the harmful effects of pesticides over wide areas of the United States. These were among the many studies drawing attention to pollution and the depletion of natural resources which in the late 1960s led to the formation of new environmental groups. The Trust was not in touch with these young populist organizations, but in the 1970s it began to share the wider public's alarm at the cumulative damage to wildlife caused by chemicals, land drainage, hedge removal and other aspects of intensive farming.

The Trust had for some time recognized that existing legislation did not give sufficient control over the 200,000 acres of land let to tenant farmers (although limited areas were excluded from tenancy agreements and managed directly). In 1984 it took the lead in promoting changes to the Agricultural Holdings Bill, which made it possible to include in tenancy agreements clauses to safeguard sites of special wildlife, archaeological or landscape importance. Since then many new tenancies have included special provisions, for example to restrict chemicals or to stipulate levels of grazing, usually in return for a reduction in rent. Although these conservation clauses are necessarily negative and cover specific areas rather than prescribing how the farm as a whole should be managed, they have had significant results at a cost to the Trust of some £100,000 or more each year.

In 1986 John Harvey was appointed to be the Trust's conservation adviser. After twenty-one years spent teaching and researching in plant ecology at Cambridge, he came with a clear idea of what he wanted to achieve: first, to bring the Trust fully into the conservation movement; second, to get more publicity for the Trust's innovative work, for example in reintroducing grazing on the Cornish cliffs, thus recreating the traditional habitats; and third, to enlarge the nature conservation staff both centrally and in the regions. With energy and determination he soon made progress towards all three objectives, though not as much as he would have liked in the recruitment of advisers in the regions.

The first archaeological site, the foundations of a Norman keep near Derby, was given to the Trust in 1899, the same year as the first nature reserve. There followed a number of much more important sites. Among the most notable were Stonehenge Down, with its fine Bronze Age barrows, bought following a national appeal in the 1920s, and the great Avebury Stone Circle, bought in 1943. Other sites were acquired incidentally as part of open spaces, for example the Iron Age earthworks on Minchinhampton Commons in Gloucestershire.

The first professional archaeologist was appointed in 1975, a year after the first botanical adviser. Until then the only systematic archaeological work was carried out on a voluntary basis by an enthusiastic amateur, Phyllis Ireland. She had worked with Sir Mortimer Wheeler (one of the Trust's honorary advisers) and in 1962 warned the Trust that a Neolithic enclosure on the Slindon estate in Sussex was in danger of being ploughed up. She offered her help and set to work compiling lists of sites for each property, region by region, using the Ordnance Survey and National Monuments records. When she retired in 1980 she had completed her self-imposed marathon.

Her place was taken by a full-time Cambridge arch-aeologist, David Thackray, who had been recruited several years earlier by the Wessex Regional Committee, to investigate sites in Wiltshire and Dorset. His new assignment was to follow up Phyllis Ireland's documentary work with a field survey and to make detailed recommendations about how each property should be managed. He started on his own in Devon and Cornwall, expecting to move on to other regions, but five years later he was promoted to take charge of a growing team of archaeologists in different parts of the country and moved to Cirencester, where the biological and gardens advisers were based.

By 1990 nearly half the country had been surveyed, but the work was slowing down as the archaeologists were asked to spend time on the surveys of vernacular buildings and historic landscapes and on investigating particular sites before building works or garden restorations took place. The need for com-pleting an archaeological survey before embarking on a new management strategy had been illustrated by a much pub-licized incident in 1982 at Wimpole near Cambridge, when a programme of drainage and cultivation to improve the grazing in the park seriously damaged the sites of a medieval village and field system that had hitherto survived. Had there been a field survey and had the ensuing recommendations been incorporated in the management plan, the mistake would have been less likely to occur.

However, the development of a wider approach, looking at whole landscapes rather than concentrating on individual sites, had a beneficial effect on the Trust's acquisition policy. Some of the most significant new properties of the 1980s were designed to consolidate and extend ownership of large areas. For example, in West Penwith, the extreme south-west peninsula of Cornwall, the coastal properties were broadened to protect farms further inland with their tiny fields, often of late prehistoric origin, enclosed by walls of massive boulders as well as the relics of the old tin mines with their distinctive if ungainly chimneys and engine houses. In Northumberland farms were bought astride Hadrian's Wall in order to safeguard its wild open setting.

In 1983 when Fountains Abbey and the Studley Royal estate in Yorkshire came on the market, the necessary grants from the National Heritage Memorial Fund and the Countryside Commission were in part conditional on archaeological surveys of the park and the eighteenth-century water gardens. Although the monastic ruins, the largest and some of the most beautiful in Britain, are in the guardianship of English Heritage, the Trust needed professional expertise to look after the estate and subsequently to investigate suitable sites for a visitor centre and car park to accommodate the increasing number of vehicles which invaded the valley and intruded on the views of the Abbey. When the visitor centre was built on the plateau above the Abbey, too contemporary in design for everyone's taste but sitting comfortably in the landscape and invisible from the valley, entirely new views of the ruins were opened up for visitors.

The continuing expansion of the Trust's estate has brought a steady increase in the number of archaeological sites; by 1990, some 7000 had been recorded out of an estimated total of 40,000 owned by the Trust.

The years 1975–85 saw the largest increase in acreage of any ten-year period in the Trust's history, an additional 160,000. Almost 40 per cent of this was accounted for by three exceptionally large estates – Leconfield Commons in the Lake District, which were transferred through the Land Fund; the Abergwesyn Commons in mid-Wales, which were purchased with the help of grants from the Countryside Commission and the National Heritage Memorial Fund; and the Kingston Lacy and Corfe Castle estates in Dorset, which were left by Ralph Bankes, whose family had owned them for 350 years.

The Bankes estate included three nature reserves which were known to be of international importance for their rare plants and reptiles, and the biological survey revealed other valuable habitats not previously known, including downland which had never been ploughed and water meadows rich in plants and birds.

The bequest was the most valuable ever to have come to the Trust in terms of capital value (£23 million) and in the range of its treasures. Kingston Lacy house contained paintings by Velasquez, Titian and other old masters; there were the ruins of Corfe Castle and the prehistoric site of Badbury Rings; and the estate included three pubs, a hotel, forty farms and 327 houses and cottages, not to mention a golf course and some quarries. Many of these were very run-down and they posed problems of financing as well as conservation. But by this time John Cripwell, the regional director, was not short of advice on archaeology, gardens, paintings, furniture and textiles as well as on nature conservation. He outlined the decisions he had to make in the Kingston Lacy park alone:

> We thought the public must have access to most of the park. Mr Bankes was justly proud and fond of his famous herd of Red Devon cattle; we must surely keep the beasts even if they were an uneconomic proposition. How many visitors could we expect in the park? Could the park be grazed by sheep as well as cattle? What would be needed in the way of car parks, drives and kiosks? Should the park be kept in hand so that the Trust has complete control of spraying, fertilizing and grazing? How many wardens and how much equipment was needed? Should the character of the park be restored to its eighteenth-century lay-out, a survey of which was stored away in the Estate Office?[6]

The estate included a popular beach at Studland Bay which was also not without problems arising from the use of a part by naturists. The continuation of this use was challenged by a resolution put down for the 1989 AGM and again in 1992 (but not formally moved on either occasion).

Another spectacular coastal property, also rich in wildlife, was Stackpole, near Pembroke, which was transferred through the Land Fund. Its eight miles of cliffs and unspoilt beaches were home to large colonies of breeding birds, its inland lakes possessed an almost unique range of aquatic vegetation and its disused stable block sheltered the greater and lesser horseshoe bats.

With the vast increase in landholding during the decade it is perhaps not surprising that, when in June 1983 the Trust was given the opportunity of buying Abergwesyn Commons, a remote moorland plateau in mid-Wales and the feeding ground for the rare red kite, there should have been some doubts. The stumbling block was not the £105,000 for the purchase price but the £200,000 calculated under the Chorley formula needed for the endowment, even though the whole area could be managed by a part-time warden with a Land Rover. The Trust dithered over these sums, small in comparison with the millions being raised at the same time for Belton and Calke, but after pressure from the Ramblers' Association and the Open Spaces Society, and with strong support from Angus Stirling, the new director-general, the commons were purchased.

Stirling took over in September 1983 at the age of fifty and became the most powerful although not the longest-serving chief executive in the Trust's history. He had been educated at Eton and Trinity College, Cambridge, and then worked for Christie's and in the City before becoming deputy secretary-general of the Arts Council under Gibson. It was at Norman's suggestion and with Gibson's support that he was brought into the Trust as deputy to Boles – and clearly as crown prince.

Tall, highly articulate both in speech and on paper, he was sensitive to people and, despite a certain reticence, attracted great loyalty from his staff. His membership of the Covent Garden Board (of which he later became chairman) and other arts organizations gave him interests and friendships beyond the Trust. A persistent taste for adventure was masked by his conventional dress and once landed him down a crack in the Arctic ice-cap, almost preventing his return to Queen Anne's Gate.

Stirling was fortunate just to have missed the first major row to shake the Trust since the Rawnsley affair fifteen years earlier. This was caused by the decision to lease a twelve-acre field to the Ministry of Defence for an RAF underground communications centre. It was set deep in the woods of the Chilterns above the village of Bradenham, where Disraeli's father had lived. The Ministry had approached the Trust officially in 1978 and had applied to the district council for planning permission two years later. The regional committee, whose chairman was Lionel Brett, the fourth Viscount Esher, architect son of the former Historic Buildings Committee chairman, did not think the matter of sufficient importance to consult head office. When Boles heard of it, he quickly informed the Executive Committee but they were advised that the Ministry of Defence could walk in at twenty-four hours' notice. Moreover they did not realize that some members would be deeply angered if inalienable land was used for a key defence installation without the issue being aired at a public enquiry or taken through the special parliamentary procedure. In July 1981 the Executive Committee agreed to the lease in principle. The Council was informed in December and gave approval in March 1982. Although a few members were not pleased at being presented with what was virtually a *fait accompli* and were initially doubtful, all but one finally agreed.

The Trust's case was that as the field itself was scenically uninteresting, was not visible from elsewhere, was already

overlooked by RAF buildings and was not visited by the public, it could not be defended on grounds of high landscape or
amenity value. For these reasons it was thought that the Trust
would be unlikely to win an appeal and would best safeguard
its interests by agreeing to a lease and imposing strict conditions on the works and the reinstatement of the land rather
than by taking action likely to result in a compulsory purchase
order. Gibson thought that there was no point in weakening
the Trust's authority by fighting a battle which could not be
won.[7] Even John Popham, an elected member of the Council
and the protagonist of many conservation battles, was sure
that this was a case for negotiating rather than fighting to the
last ditch.

Meanwhile Trust members first heard of the proposed bun-
ker in the summer of 1981, alerted by Trevor Hussey, who had
lived in Bradenham all his life. Professor George Hutchinson of
Southampton University, a prominent member of the Campaign for Nuclear Disarmament and a life member of the Trust,
wrote to object and to ask what steps had been taken to inform
members. In reply Julian Prideaux, the regional director (later
chief agent), added to his explanations about the site the information that the Ministry of Defence had assured the Trust that
the project was extremely important and there was no alternative site.

Letters of protest followed from other members, including
Professor S. A. Urry, whose wife Audrey became the organizer
of the campaign against Bradenham. She was concerned about
'the secrecy involved; how late in the proceedings it was before
Council were informed about the matter; the misleading way
the project was described to Council and members – as if it
were a very minor affair on an insignificant piece of land'.[8] The
leaders of the protest group were very different from those
who had supported Rawnsley: most were keen supporters of
the Trust, many were also involved in the peace movement,
some lived locally. A small group met early in 1982 and

decided to seek publicity for the Bradenham lease and urge members to protest. Encouraged by the reaction, they decided to requisition an extraordinary general meeting and within a few weeks had obtained more than 3000 signatures, most of which came in response to a letter from Mrs Urry published in the *Friend* (the Quaker weekly journal). Support also came from Lord Beaumont, a Liberal peer, who thought that 'the affair had all the marks of a gentleman's agreement arrived at over a table at Brooks's or whatever club the higher echelons of the War Office inhabited'.[9] He offered to act as plaintiff in challenging the legality of the lease in the High Court, where he was represented free of charge by John MacDonald QC. The case did not come on until after the EGM and was lost, leaving the protest group £16,000 to raise for the costs.

In requisitioning the meeting Mrs Urry, Professor Hutchinson and others put down two resolutions. The first deplored the decision to lease the Bradenham site to the Ministry of Defence and the failure to provide adequate information to members, and the second called for the resignation of Gibson, Boles and Esher. In support they pointed out that the site was within both the Green Belt and an Area of Outstanding Natural Beauty and that the wood through which a temporary access road would be built was a Grade I Site of Special Scientific Interest. The damage would extend beyond the twelve acres to a further twenty-seven acres where soil would be dumped. They argued that in not opposing the development of inalienable land the Trust was setting a precedent which would put other properties at risk and might deter future benefactors. They also complained that the Trust had not taken the opportunity to consult members during the protracted period when the project was under discussion and had brushed aside their requests for information.

The Trust circulated at its own expense the opposition's leaflet in accordance with the Benson Committee's Constitutional changes, and also sent out its own leaflet headed 'An

attack on the way the National Trust is run will be made at the
Extraordinary General Meeting and the Ordinary General
Meeting at Wembley Conference Centre on Saturday 6
November 1982'. This stated that 'The Trust always fights
whenever land of real quality is threatened' and for the first
time mentioned the political undertones of the dispute:

> Because it is an underground headquarters for the RAF
> which is required, this issue has become entangled in the
> minds of some members with arguments concerned with
> national defence, nuclear disarmament and so forth. It is not
> the Trust's job to involve itself with these or any other
> matters other than the simple issue of how best in any situa-
> tion to protect its property.[10]

A six months' picket by the Wycombe Peace Council, followed
by marches and vigils, helped to keep the issue in the public
eye.

Both the Trust and the protest group had written to the
members' centres urging them to attend the meeting. More
than 2000 members did so and the proceedings, although
interrupted by a good deal of heckling, were much less acrimonious than the Rawnsley ones had been. Gibson handled
the meeting with authority, giving both sides plenty of time to
express their views. Mrs Urry set the tone by moving the
resolution in a low-key, reasoned manner, and Sir Richard
Acland, who made the most effective speech of the day, was
free from bitterness. He did not believe that the Trust had not
known that the question of whether to fight the proposals for
Bradenham had, for the first time, raised ethical and political
principles; he thought that many members would have preferred the Trust to have been democratically overborne by
Parliament (which would certainly have granted the concessions) rather than find out that a semi-secretive agreement had
been made. Sir Marcus Worsley, on behalf of the Council,
argued that, as the Trust owned vast acreages of land, it would

be wrong to freeze every acre irrespective of merit and if it tried to do so the privilege of inalienability would be removed. There were some 500 examples where, to some degree, inalienable land had been used for public purposes. The Council secured a vote of 169,924 in its favour to 26,619 against, though it lost the vote among those present in the hall.

After the meeting the chairman of the Dorset Centre wrote to Boles criticizing the Trust's tactics in not seeking a public enquiry, which 'would have drawn the force of the attack and adverse publicity' and noting that some members were seeking involvement in the management of the Trust. In reply Boles reiterated his view that the Ministry of Defence would immediately have invoked compulsory purchase procedures but conceded that 'We certainly must plead guilty to misjudging the reaction of a significant number of perfectly responsible members, although I think it is fair to point out that only 2.5 per cent voted against the Council (16 per cent of those voting). Because it shares your concern, as I do, the Council has set up a sub-committee.'[11]

As after the Rawnsley extraordinary general meeting, no time was lost in taking action. The Council, having been placed in a difficult position, determined that in future applications for inalienable land should in all cases of doubt be referred to it for decision. A committee was set up under the chairmanship of John Arkell, who had represented the CPRE on the Council since 1970 and had applied for the post of secretary to the Trust after the war, to review: (a) the relationship between the members and the Council and management of the Trust; (b) the procedure for dealing with all aspects of inalienability; (c) any resulting changes in structure or procedures.

The Committee, which included four elected members of the Council and was serviced by a management group headed by Stirling, reported within five months. It made a number of proposals relating to how the members – well over one million by 1984 – could be better informed (an extra number of the magazine, more open days, more opportunities for voluntary work) but emphasized that the first priority must be the effective management and preservation of the Trust's properties. The Committee was particularly critical of the narrow social base of the regional committees and recommended that vacancies should be advertised in the regional newsletters and an appointments panel set up to scrutinize their membership. Mrs Urry and Professor Hutchinson gave evidence to the Arkell Committee and their collaborators continued afterwards as an active ginger group, meeting from time to time and circulating their own newsletter.

Stirling had already been at the Trust for some four years when he succeeded Boles, time enough to get to know the staff and the properties, but not too long for the sharpness of his first impressions to be blunted. He was impressed by the commitment of the staff and the quality of their work; but within a week of his arrival he felt that the Trust was too inward-looking, was not conscious of its role in the outside world and was not sufficiently interested in external opinion. Perhaps the fact that only a narrow section of external opinion was recognized (the *Daily Telegraph* and *Country Life* were taken but not the *Guardian*) was one reason why the outcry about Bradenham had caused such surprise. Stirling was also struck by the deferential attitude and the old-fashioned feeling of 'gentlemen' versus 'players' which he felt was not conducive to the best relationships. He was doubtful whether he would ever succeed in changing such a conservative organization but Fawcett, who had himself introduced so many innovations, was confident that he would. Stirling set himself three objectives: to plan ahead in a more organized way, setting priorities for the next few years; to continue the process of making the management more professional; and to show that a concern for people inspired the preservation of the properties as much as it had in the founders' day.

Stirling believed that the Trust could not be efficiently run until there was a clear distinction between the responsibilities of committees and staff, a problem that had not been addressed by the Benson Committee. On his advice, the Council asked Simon Hornby, the chairman of W. H. Smith and a long-standing member of both the Executive and the Council, to conduct a review of the relations between committees and staff. With him were Len Clark, Ladislas Rice, a member of the Finance Committee, and Roger Chorley, together with Stirling and the Trust's secretary, Ivor Blomfield.

Hornby kept his committee hard at work and wrote the report himself. It marked the completion of the Trust's transformation from a small charity in which the Committee members both made and carried out the decisions (as Octavia Hill had for the Priest's House at Alfriston) into an organization which in 1985 had a staff of 1750 managing a large and complex estate. The key recommendation was that the Executive Committee should act as a board of directors and that the director-general should be clearly established as the chief executive to whom all the staff, at regional as well as national level, would report. More authority for decision-making would be delegated to staff – for example, on the acquisition of small properties – and the principal committees would meet six rather than eleven times a year. In accordance with the main recommendation, regional committees would cease to exercise management functions and would have a purely supervisory role, a change which the regional chairmen were prepared to accept. As a corollary, the Hornby Committee emphasized that clearly prescribed policies had to be laid down by the Executive Committee within which the management must work; and the Council, acting in their position of trustees, had to be clearly responsible for the general principles guiding the Trust.

The report was presented in 1984 and resulted in a decisive shift in the balance of power between the voluntary committees and salaried management. This was due not only to the increased delegation of decision-making but also to the elapse

of two months between Executive Committee meetings. Informal relationships became more important between the chairman of the Trust and the director-general, between the
chairman of the Properties Committee and the chief agent and
historic-buildings secretary, and between the chairman of the
Finance Committee and the director of finance. Another consequence of the report was that regional directors were made
accountable for all aspects of management, which brought to
an end – or almost did – the regular overspending on large
projects.

Stirling's desire to widen the Trust's horizons had already
received encouragement from across the Atlantic with the
establishment of the Royal Oak Foundation in 1973. This had
been set up by a group of eminent Americans to support the
Trust and give financial assistance for special heritage projects.
Particularly in the early years donations were made for a
variety of buildings and sites, including Battle Abbey and the
Grand National course at Aintree, which had nothing to do
with the Trust, but as time went on most donations were
concentrated on the Trust, which received a total of $2 million
during the years 1973–93. Apart from grants for Washington
Old Hall, the White Cliffs of Dover and other high-profile
properties, which were partially offset by payments from the
Trust to help build up the Royal Oak Foundation's membership, the close links between the two organizations were valuable in promoting regular contact between people interested in
conservation in the two countries. The Royal Oak Foundation
Board held its summer meeting in London and its members
included the chairman and director-general of the Trust. It also
financed short study courses for Americans in England; in the
reverse direction the Trust supplied a number of people to
lecture in America.

In 1985–6 the 'Treasure Houses of Britain' exhibition in
Washington focused attention on the Foundation and also on
the Trust, which was the largest single lender and had

seconded Jackson-Stops to be guest curator and to prepare the sumptuous catalogue.[12] Damaris Horan was appointed executive director of the Foundation and initiated a long-term development programme and a more ambitious series of lectures which have widened its base of support and raised its profile in cities across the States.

The Trust was slower to develop European than transatlantic links. As a member of Europa Nostra and of the International Castles Institute from the 1970s, the Trust was represented at those organizations' conferences, but not until the later 1980s did the Trust make its voice heard directly in the European Community.

CHAPTER FIFTEEN

A Personal Perspective
by Jennifer Jenkins

I WAS SURPRISED, but flattered and excited, when in the autumn of 1983 Pat Gibson asked me whether I would join the Trust's Executive Committee and be a candidate for the chairmanship when he retired. He was careful to add that the succession was not in his gift and there would be other candidates. My ten years as chairman of the Historic Buildings Council for England, preceded by a period as a part-time secretary of the Ancient Monuments Society, had given me some experience in fields linked to the Trust's work. I knew many of its historic houses and, as a walker, I could usually tell when I was on Trust land from the standard of maintenance and the quality of any new buildings or car parks. But I had no knowledge of farming, forestry or estate management and was only a very amateur gardener, so my qualifications for the job were patchy. However, I was appointed to the Executive Committee in January 1984, and in March 1986 became the first chairman since Robert Hunter who was neither from a land-owning family nor country bred.

I cannot easily judge the years of my chairmanship. But nor can I leave them out of what purports to be a history of the Trust's hundred years. All I can do is to attempt a brief account of how I saw the modern Trust working and what were some of the more significant events of those years; and then to show how I see the Trust fitting into the national and international scene.

I was fortunate to preside over five buoyant years when membership rose from 1.3 million to over 2 million and annual income from £39 million to £71 million (an increase in real terms of just over 50 per cent). The Trust's landholding also increased substantially, although less than it had during the previous five years. Stowe, the outstanding monument to English landscape gardening, came to the Trust, but no major country house apart from Kedleston, which had been agreed before my chairmanship (although considerable efforts were still necessary to finance the preservation of the mansion and the furniture which had been made for it 225 years before). But negotiations were under way for several special additions to the Trust's properties, such as A La Ronde in Devon, a semi-folly with much of its decoration made from feathers and shells, and Chastleton House in Oxfordshire, almost unaltered since it had been built early in the seventeenth century. Despite my efforts to secure Cragside, Canons Ashby and Calke Abbey for the Trust when I was at the HBC, I was not sorry that there was no continuing flow of large country houses. Everywhere I went I was asked about the Trust's great houses, seldom about the coast or countryside or even the Lake District. One of the aims I set myself was somewhat to redress this imbalance. This was not because I was uninterested in the houses, but – at least in part – because I thought that threats to the countryside had become more pressing than threats to historic houses.

I was also fortunate to preside over a deeply committed and by this time highly professional staff numbering about 2000. Angus Stirling was well supported by his deputy, Ivor Blomfield, whose diplomatic skills in person and on paper could be relied on patiently to sort out any difficulties within the Trust or with outside organizations. A complete contrast was Julian Prideaux, the chief agent, a man of great practical ability, whose dynamic energy got things done in the shortest possible time. From 1988 the Trust's finances were in the hands of Ian

Bollom, who, unusually for an accountant, always sought to find a positive solution.

I found the Trust very different from any organization with which I had previously been involved. It was more hierarchical and its staff were therefore rather disconcerted by my practice of talking to whoever was doing a particular job, whatever his or her status. It was also exceptionally male-orientated. Having been chosen as its chairman I could hardly say that its outlook was male-chauvinist, but only two regions had women chairmen, there was no woman on the Management Board and few among the core of its middle management – the land agents and historic-buildings representatives. The first senior women executives were appointed during my time (although not as a result of my intervention): Anne Roberts, a twenty-seven-year-old Scot, who became director of enterprises, and Susan Woodeson, who became director of personnel. Shortly afterwards, Tiffany Hunt became the first woman regional director, having come up through the Public Affairs Department.

The astonishing growth in the number of members showed that the recruiting techniques were working well. But arrangements for servicing members had not kept pace. Staff and committees were accustomed to concentrating on the management of properties and it was some time before the needs of members were fully considered – long after subscriptions had become a more important source of revenue than rents.

My mind had been concentrated on how best to handle a large membership during my eleven years as chairman of the Consumers' Association. When I arrived at the Trust it was apparent that the central computer needed replacement: it was inflexible and could hold only a limited amount of information; it could not be used for monitoring renewals or for fund-raising. In 1987 Brian Lang, who had been secretary of the National Heritage Memorial Fund, was appointed to take charge of public relations, and he soon set in hand the commissioning of a new computer system. However, like most advanced systems, this took several years to plan and, also far from uniquely, the first attempt failed, though without involving the Trust in much loss other than of time.

The Arkell Report had led to various measures for keeping members more in touch through regional newsletters and information boards at properties. The magazine was published a little more frequently and was greatly improved in lay-out and content. But the readers' reaction was not closely or regularly followed up and some services were still run more for the satisfaction of the management than for the convenience of members. For example, until 1988 there was no national brochure giving particulars of holiday cottages let by the Trust (each region had to be contacted separately) and until 1994 there was no central booking arrangement.

The retailing and catering enterprises under successive directors with professional retailing experience were already more attuned to meeting consumer demands, and this was underpinned by Anne Roberts when she was appointed in 1988. She brought a new emphasis on profitability and introduced young designers and a wider range of products, including up-market items ranging from porcelain birds to furnishing fabrics. There were also two entirely new ventures – the Frizzell Homeguardian Insurance Scheme and the Midland Bank affinity card. Such financial fringe activities may sound incongruous to the central purpose of the Trust but they have been invaluable in providing money for its core conservation role. The first produced over £1 million for the Trust in the first four years, and the second almost as much.

Like most large organizations the Trust has a fairly complicated committee structure. And as in most voluntary bodies the essential purpose of it is, or should be, to strike a satisfactory balance between the freedom of the professional staff to manage and the ability of the voluntary officers (though the chairman was nearer full time than half time) to receive enough information and to exercise enough policy control to discharge their responsibility for the efficient and sensitive running of the organization.

The committee pattern tried to meet these objectives. Every two months there was a 'meetings' week – the Properties Committee on Tuesday afternoon, the Finance Committee on Wednesday morning and the Executive Committee on Thursday morning. I attended all of them, although I presided over only the Executive Committee. Twice a year a meeting of the full Council followed the Executive meeting on a Thursday afternoon. Again I presided over the Council. My first such week was in May 1986, and a formidable concentration of effort and stamina was required. In the two other quarters the Council also met, but not on the same day or in the same week as the Executive. Since the Benson Report, the management of the Trust's affairs had increasingly passed to the salaried staff, but the voluntary committees, with their advisory panels, still played a crucial role and the Council had become more influential.

The Properties Committee had some twenty members, who included not only John Cornforth, the architectural historian, Peter Fowler, a leading professor of archaeology, and experts on other aspects of the Trust's work, but also David Gentleman, the landscape painter, and John Julius Norwich, the writer. It was chaired by Marcus Worsley, for many years chairman of the Yorkshire Regional Committee and a long-standing member of the Executive Committee and Council. His great interest was forestry (he had won a number of awards for his Yorkshire estate at Hovingham), but he also knew a lot about buildings and farming. With the natural relaxed authority of a large man of substance, he steered the meetings through the heavy agenda at a brisk pace, but with a light touch. The main business of the Committee was to oversee the management of the Trust's estate, and for this reason it would usually discuss a particular aspect of the Trust's work – industrial archaeology at my first meeting – before going on to consider whether proposed acquisitions met the Trust's rigorous standards.

Next morning there was the Finance Committee, smaller in number and chaired by Nicholas Baring, at that time deputy chairman of the family bank. He had been brought on to the Executive Committee as a young man by his uncle Lord Antrim and had then been elected to the Council. Under his guidance the Finance Committee went through a regular schedule of accounts, salary reviews and budgets as well as such longer-term matters as financial planning and legacies, before considering the funding of new acquisitions. Financially the Trust has to walk on a tightrope – its properties once acquired and declared inalienable cannot be disposed of but cannot be securely provided for in advance to more than a limited extent. It was reassuring to know that a man of Baring's patience and acumen was supervising this delicate balancing operation.

Marcus Worsley and Nicholas Baring played an influential part on the thirty-three-strong Executive Committee, as did several of the regional chairmen, notably Spencer Crookenden from the north-west. Crookenden saw clearly the connection between conservation and people, at his last meeting making a memorable and amusing plea for the historic houses not to become fossilized but to make room for contemporary even if eccentric art and design. This was in fact already happening through the Trust's Foundation for Art, which had been set up in 1986 to commission paintings recording the houses and gardens and the personalities associated with them.

At my first meeting the Committee approved three disparate acquisitions, each of which foreshadowed others of their sort in the next few years: cliffland in Durham, the first purchase of coastline in that county; Sheringham Park in Norfolk, considered by Repton to be his finest design; and an area of marshland in the Broads of that same county, important for its natural life. Among the other items on the agenda were the arrangements for trying to deal with the summer solstice invasion at Stonehenge (a problem little nearer solution eight years later) and the disposal of land owned by the water authorities after privatization.

By the autumn of 1986 I had come to feel that the gap of two months between Executive Committee meetings was too long to be without up-to-date financial information or a systematic report on major developments or other matters which were sensitive from a parliamentary or press point of view. Without a monthly meeting I felt that the balance of power had swung too far from the elected chairmen, who in the last resort had to answer to the Council and the annual general meeting. And so from the autumn of 1986 Marcus Worsley, Nicholas Baring and I had a monthly meeting with Angus Stirling, Ivor Blomfield and the secretary (at first David Beeton, ex-chief executive of Bath City Council, and later David Pullen, ex-solicitor of Buckinghamshire County Council). This steering group provided a useful means of keeping in touch with current issues and of preparing for the larger meetings.

The summer Council meeting was held at a Trust property out of London. My first was at Ashridge, the estate near Berkhamstead, for which G. M. Trevelyan had raised a public appeal. Having disposed of the business, we all rambled through the woods at Bradenham to look at the work on the RAF communications centre which had been one of the most explosive issues under the previous chairman. It had provided a salutary warning of the pitfalls lying in the path of any big voluntary organization with a vigorous and potentially critical membership and a natural resonance in the national press.

The twenty-five nominated members of the Council between them had a wide range of experience, from the Duke of Atholl of the Scottish Trust and Professor Guenter Treitel of the British Museum to Hermione Hobhouse of the Victorian Society and Nicholas Biggs of the British Trust for Conservation Volunteers. The elected members included Marcus Worsley, Nicholas

Baring, six regional chairmen, Simon Hornby and John Smith; most of the others were connected with local associations or with other conservation organizations and three had worked as young volunteers.

The Council had tended to have a stable membership, some would have complained too stable, of elected and nominated members alike, but in my second year (1987) there was a sudden increase in the number of candidates seeking election, and this resulted in a marked shift in the type of members coming on for the first time – fewer regional chairmen and more of those active in local associations. Both before and after this change, however, the Council showed itself to be an admirable body for giving general direction to the Trust and deciding its attitude on matters of public policy. Most members are careful to consider the interests of the Trust rather than to advance their own pet objectives or those of their organizations, particularly important when the often contentious resolutions for the annual general meeting were discussed. These AGMs, my first two in Exeter and Buxton and the third in London, were far from resembling routine meetings of companies with good dividends and respectful shareholders, at which re-elections and votes of thanks are mumbled through. They were well attended, often lively. And presiding over them was the greatest annual test of nerve and knowledge of the business of the Trust which the chairman had to face.

There were two very different AGM issues which remain strongly in my mind. The first, relating to the Lake District, was entirely germane to the running of the Trust, therefore appropriate to the AGM, and although raised in a fairly abrasive manner was in the outcome highly beneficial. The second, the hunting issue, was essentially external to the Trust and in my view diverted – and continues to divert – its energy rather than focusing its attention on its proper purposes.

In 1986 Dr John Wilks, an Oxford don and a regular visitor to the Lake District, put down a resolution calling on the Trust to raise an additional £1.5 million a year to overcome the backlog of work in the Lake District, and in his supporting statement criticized the Trust for failing to provide adequate conservation over a long period and more recently to pursue the Lake District appeal with vigour. In its reply the Council pointed out that the labour force had been considerably increased over the previous two years and progress made on repairing footpaths, walls and buildings, but recognized that the appeal should be vigorously promoted. This was the first appeal for continuing countryside maintenance rather than for acquisitions, and it was not until an exceptionally effective fund-raiser, Bernie McDermott, was appointed in May 1986 that it really took off: £2 million net was then raised in four years, well over the target figure. An institution never finds it agreeable to be pilloried, and Dr Wilks did not temper his criticism with emollience, but he undoubtedly performed a service in drawing attention to the needs of the Lake District.

A side benefit was that the appeal led directly to the Trust's first survey of an entire landscape. Studies had already been completed of individual elements – archaeological sites, farms, woodlands and ecology – but not of how these fitted into the overall pattern of walls, hedges and footpaths. The expanded programme of work involved constant decisions on priorities. Which walls should be repaired? If footpaths had to be restored, would the new route disturb rare plants or ancient earthworks? Susan Denyer, the historic-buildings representative, realized that such questions could not be adequately answered without a grasp of the landscape's history. In 1987 she recruited a team consisting of two surveyors, a cartographer, an archivist and a historian to record all the man-made features in Great Langdale, to search out documentary references and to interview the older farmers and farmworkers. Once their work had established the pattern of development they moved on to other valleys.

The most powerful pressure groups seeking to influence

the Trust in the 1980s were the animal-rights organizations, which were intent on securing the abolition of hunting on Trust land. The issue had first been raised at annual general meetings as long ago as 1934 and 1936 (the League Against Cruel Sports had been founded in 1924). Between then and 1988, however, it produced more subterranean rumbling than above-ground explosives. In that year it resurfaced powerfully but not unexpectedly. As Sir Keith Thomas' study of men's changing attitudes to the natural world had pointed out:

> The early modern period had (thus) generated feelings which would make it increasingly hard for men to come to terms with the uncompromising methods by which the dominance of their species had been secured. On the one hand they saw an incalculable increase in the comfort and physical well-being or welfare of human beings; on the other hand they perceived a ruthless exploitation of other forms of animate life. There was thus a growing conflict between the new sensibilities and the material foundations of human society. A mixture of compromise and concealment has so far prevented this conflict from having to be fully resolved. But the issue cannot be completely evaded and it can be relied upon to recur.[1]

Recur it did at the 1988 annual general meeting, when Mr Paul Sheldon moved a resolution to prohibit all fox, deer and hare hunting with hounds on Trust land, arguing that, as civilization moved on, such practices became unacceptable. This resolution failed, but in 1990 the crunch came when two separate resolutions were moved, the first to prohibit the hunting of deer and the second of fox, hare or mink. The first was carried by 69,000 votes to 64,000 and the second was defeated by a similar margin; only just over 7 per cent of members exercised their votes. This time the arguments were based not only on cruelty but also on how the Trust's purposes and by-laws should be interpreted.

The Trust on both occasions reaffirmed its policy as being to allow hunting where it is rooted in local tradition, does not contravene the wishes of those who gave the land and is not harmful to nature conservation, public recreation or tenants' rights. My own instinctive feelings were probably somewhat hostile to hunting, although tempered by a strong general inclination to let people do what they wanted to provided it did not result in some demonstrable social evil. And the demonstrable quality of the evil was in this case far from clear. There was a good deal of evidence that some of the various species were at least as well off (if survival is regarded as being better off than extinction) with than without hunting. However, all these propositions were highly arguable. What was indisputable was that the essential purpose of the National Trust was not to be an advance column of the anti-hunting lobby. To hunt or not to hunt was an issue which should be decided on a national basis in Parliament, on a free vote as, say, with abortion, and not on an estate-by-estate or even county-by-county basis and over the bleeding body of the National Trust. The hijacking by special-interest groups of institutions set up for entirely different purposes is a real danger.

Banning of deer hunting raised the special problem of how the deer could be conserved without the co-operation of the farmers through the hunt. At my suggestion the Council decided to set up a working party to study the management of red deer on Exmoor and the Quantocks, this having the advantage of allowing time for Private Members' Bills to be considered in the House of Commons both as it was before and again after the 1992 general election.

The anti-hunting resolution came about in an entirely different way from those initiated by the Rawnsley and Bradenham groups. These two mobilized their support from scratch, relied on their own slender financial resources and directed their attack to the Trust alone. In contrast, the move to ban hunting was organized by well-established and well-funded groups as part of a general national campaign, directed ultimately to Parliament. The victory on deer hunting was followed by demands that resolutions carried at a general meeting should be binding on the Council and that postal voting should replace proxy voting. In December 1991 the Council, at the suggestion of Roger Chorley, who would follow me as chairman of the Trust, referred these questions to the former Law Lord, Lord Oliver of Aylmerton, as part of a review of the working of the Trust's constitution. He categorically rejected both of the proposed changes while suggesting a number of minor amendments. Commenting on the hunting dilemma, Lord Oliver said, 'The fact is that there are two entrenched and irreconcilable views regarding field sports, both sincerely held by their adherents and both capable of being supported by rational and respectable arguments. I do not, however, think that either has much to do with the objects for which the Trust was founded.'[2]

Before I had heard of Dr Wilks' resolution I had arranged that the first of my visits outside London (the most enjoyable of a chairman's duties) should be to the Lake District, since this was where nearly a quarter of the Trust's estate was concentrated. The large land ownership brought with it special responsibilities to the local community as a provider of jobs and housing, and I was anxious to have an opportunity of discussing these questions with the regional committee and the National Park Special Planning Board. Despite occasional differences of opinion, it was clear that the two organizations worked well together, and this was confirmed when the National Park chairman told me that he would like to see the Trust's ownership further extended.

One of the problems both had to tackle was the great scars on the fells caused by the thousands of walkers. Neil Allinson, who enjoyed the rather grandiose title of upland access adviser, had already led me up one of the 'pitched' paths, which teams of workers were building from blocks of stone to repair the badly eroded slopes. The idea had come to him when he had discovered one of the old packhorse tracks across the mountains and had realized that, although an enormous labour to construct, they would withstand any number of feet. Allinson had left coal-mining nearly twenty years earlier to become the first of what by 1990 were twenty-seven wardens in the Lake District alone. During that time the wardens had become a new breed of countryside manager in all regions of the Trust. Mostly drawn from farming or with a science degree, they provide a link with the public and are responsible for nature conservation as well as for regular maintenance of paths, drainage and field boundaries.

Occasionally one of the tenant farmers, on whom the Trust depends for looking after its agricultural land, doubles up as a part-time warden. One of those I went to see was in the course of converting his holding near Morecambe Bay to organic methods and needed the extra income. Getting a living from marginal farms in the harsh Cumbrian climate is not easy and for another farmer whom I met on a later visit was made still more difficult by the after-effects of Chernobyl and the consequent need for his sheep to be regularly checked. I had heard the news of the disaster while I was having a sandwich lunch with members of the regional committee and tenant farmers on the Ysbyty estate in North Wales, but none of us had imagined that the fall-out would reach as far as Great Britain or would be so long-lasting in its effect.

Enterprise Neptune continued to flourish, raising well over £1 million each year in the late 1980s, and I gave it all the encouragement I could. It is an unavoidable paradox that while the Trust is in some ways part of the 'leisure industry' (not the most attractive of phrases) and an encourager of tourism, it has also to fight a running battle against the depredations of too many leisure enterprises and too many tourists. In 1987 twenty miles of coast came to the Trust, including two major gifts,

Cape Cornwall near Lands End and a spectacular stretch at the other end of England, unspoilt even though in an industrial rather than a rural county, and recreationally the more valuable for that reason. This was near South Shields and a gift from the South Tyneside Borough Council. Then in 1988 the Trust's 500th mile was reached with the acquisition of cliffland a little further south in County Durham, which had been reclaimed from mining spoil and was presented by British Coal at a ceremony in the appropriate but for the Trust the unusual location of the Easington Working Men's Institute. Once the blackened beaches have been cleaned by the tide and the gaunt winding gear dismantled, the configuration will not be dissimilar to parts of Devon.

In 1990 the Duke of Edinburgh launched Enterprise Neptune's Silver Jubilee Appeal, which raised £2 million with the help of the seventy-year-old Robert Steel's walk round the coast of England, planned with military precision and accomplished in 100 days. The most memorable acquisitions of the year were the wide sweep of farmland leading up to the ruins of Dunstanburgh Castle on its rocky Northumberland headland, and the Whistling Sands, sheltering below the windswept Llyn Peninsula in North Wales; many small pieces were fitted into the jigsaw on the coast of Devon and Cornwall.

Inland several very different landscapes were acquired. One was the magnificent Fulking escarpment of chalk downland north-west of Brighton, where on one side agricultural improvement threatened the plants and butterflies and on the other suburban housing had been creeping inexorably upwards. Much more remote but nonetheless a potential target for caravans and marinas was the Crom estate on Upper Lough Erne in the south-west corner of Ulster. In a pivotal position linking the Erne and Shannon waterways, it needs to attract visitors to occupy the holiday cottages within the massive old farmyard and the ornamental park buildings, but it will not be easy to get the right numbers while retaining the peaceful atmosphere of this haunting landscape of water and woodlands, waterlilies and dragonflies.

A contrasting landscape, once far from the madding crowd, now thronged with cars at weekends and holidays, came with 6000 acres in Upper Wharfedale, Yorkshire. Here the Trust's usual task of piecing together the component parts had already been accomplished by the donors, two Watson brothers who had themselves lived in urban Bradford all their lives while dedicatedly assembling the estate, farm by farm and wood by wood, over fifty years. On a foggy day in late November 1989 the surviving brother, as individual in his choice of transport as he had been in his expenditure pattern, drove up the snow-covered road on his high-powered motorcycle, clad in black leather, to present me with the deeds of this typical Yorkshire dale, crisscrossed with stone walls, grey farmhouses nestling in sheltered corners, the land rising to limestone pastures and moorland above.

I was also anxious to exploit the Trust's strength in vernacular buildings, where the National Trust for Scotland had moved somewhat ahead of us, and to develop interests beyond the broad acres and sylvan parkland of pre-Industrial Revolution Britain. Until Martin Drury became historic-buildings secretary, the Trust tended to assume that only country mansions were worthy of preservation; little attention was paid to the hundreds of farms and cottages scattered across the agricultural estates. Drury realized that if proper care was to be taken of these small buildings the first step was to find out more about them, and in 1982 he initiated a survey of all the Trust's vernacular buildings. Designed as a tool for the land agents and building managers, rather than as an academic exercise, it comprises a detailed inventory, with photographs, of all dwellings and farm buildings, and in the Lake District is supplemented by accounts of the families who had lived there.

The survey has revealed sharp differences in the patterns of building between different parts of the country and even between neighbouring valleys. It has also discovered some rare survivals and, above all, it has created a new interest in vernacular buildings and in the use of lime and other traditional materials for their repair. Several cottages have been opened to the public – a typical agricultural worker's dwelling in Surrey, another on the Gunby estate in Lincolnshire, a fenman's home in Cambridgeshire, and a medieval cob house in Devon where wall paintings of St Andrew have been uncovered. In none of these had the tenants left any furniture. It was rewarding to find a small eighteenth-century farmhouse on the Ysbyty estate in North Wales, quite unmodernized (no mains water, electricity or drainage), and still containing its Victorian furniture – oak settle and dressers, brass beds, grandfather clocks and a piano.

The Trust, I thought, ought to have a role in giving the public, and particularly perhaps children with an awakening interest in the 'olden days' which often begins for them about the time of their grandparents' youth, an opportunity of seeing how townspeople as well as agricultural workers used to live. No opportunity occurred during my time, but soon afterwards the Trust was left the contents of a semi-detached house in Worksop, Nottinghamshire, which had belonged to a family of seed merchants and still had all its 1930s furniture, its drawers and cupboards stuffed with belongings of the time (a smaller and more recent version of Calke Abbey). The Trust decided to use the accompanying legacy to buy the house and see if the public was interested. They were: 14,000 visitors came in the first year.

Another idea which took longer to bear fruit than I had hoped was the acquisition of a Modern Movement house which would help bring the Trust's collection of buildings up to date. The opportunity occurred soon after I retired when the Trust heard of the prospective sale of a more sophisticated house in Hampstead, built just before the Second World War by a Hungarian refugee architect, Erno Goldfinger, in an uncompromising contemporary style, and including the furnishings he had designed as well as the works of art he had collected. The Goldfinger family agreed to offer the house in lieu of tax; lengthy negotiations with the Treasury were still continuing as I write.

South Wales was in the late 1980s the only region where the Trust owned no large country house with historic contents. Until recently there was less public interest there in the architectural heritage than in England and no active building-preservation trust. Peter Mitchell, the regional director, took up with enthusiasm my idea of trying to set up an organization modelled on Hearth in Northern Ireland, which is supported by the Trust and the Ulster Architectural Heritage Society and has successfully restored over a hundred modest buildings, thus providing a catalyst for the revival of a number of small towns and villages. The project was more difficult to realize than I had expected, the initial partnership with a housing association proving unsuccessful, and it was not until 1993 that a buildings-preservation trust – Hendref – was created. Meanwhile Mitchell had taken direct action to demonstrate that traditional Welsh buildings which had been abandoned could be restored and reused. He assembled the funds to buy a late-eighteenth-century terrace of small houses, a shop and an inn flanked by a Baptist chapel and a vestry, in Cwmdu, in mid-Wales. The terrace has been rescued from dereliction, and the village revitalized with the reopening of the shop and the inn and the reoccupation of the dwellings.

Not all new policy directions were planned. Crucial to changing the Trust's attitude to its urban properties was the experience of Sutton House in Hackney, one of the few London houses to survive from Tudor times. It had been acquired in the 1930s, repaired with funds raised by public appeal, and then let, most recently to a trade union. In 1982 the lease was unexpectedly given up and the house lay empty for

323

five years while efforts to persuade the borough council to take it on as a museum or to find a suitable local organization to use it were unsuccessfully pursued. Meanwhile the house had been occupied by squatters, and there had been thefts of panelling and fireplaces. To secure the fabric the Trust had, in the absence of a better solution, decided to lease the house to a developer for conversion to flats.

When this proposal became known there was an explosion of local anger, and a Save Sutton House campaign was organized, calling on the Trust to restore the house as a museum. The campaign presented a detailed case against the proposed alterations preceded by a statement putting the objections in context: 'In an area which has 20,000-plus unemployment and has been termed "Britain's poorest borough", it is not surprising that there is indignation when it is perceived that Sutton House is in danger of being handed over to a private developer for yet another highly priced conversion.' When Angus Stirling and I went to look at the plans on site in July 1987, we found the developer and also Mike Gray, the leader of the campaign, who explained his ideas of how the house might be used. We were persuaded that his alternative should be seriously examined, for to divide this sizeable, but by no means vast, house into self-contained units would have destroyed its historic character and would have made regular public access difficult.

A fortnight later the Executive Committee decided that the Save Sutton House group should be asked to present precise proposals and that an archaeological survey should be commissioned. From that time the Trust and Mike Gray worked closely together to bring about the restoration of the house for educational and community uses and to find means of raising long-term income from a variety of sources – a shop, a café-bar, fees from letting the main rooms and rents from offices in the Edwardian wing (taken, appropriately, by the Young National Trust Theatre and the Early Music Centre). When the house was finally reopened in February 1994 it offered a new resource for this deprived East End borough; the Trust had contributed £1 million to the costs and was halfway to raising a further £700,000 by public appeal.

Another London property which the Trust took steps to make more accessible to the public was Morden Hall Park. The 124 acres of grassland are bisected by the River Wandle, in outer southern suburbia, not far from Watermeads, a nature reserve and the object of Octavia Hill's last fund-raising effort. The initiative came from the leader of the Merton Borough Council, who in January 1989 invited me to discuss proposals for a riverside path leading to a group of historic buildings restored by Sainsbury's, in expiation, as it were, of having imposed a large new shopping centre on the edge of the site. The borough council was already co-operating with the Trust in running Morden Hall's old snuff mill as an environmental centre for schools, but felt that the park was run down (as it was) and that more could be made of the old walled garden and other neglected areas. Julian Prideaux and the land agent who accompanied me set things in motion and by 1994 the bridges and paths in the park had been repaired, a tea room, shop and garden centre had been opened in the walled garden, a site had been provided for a children's farm and there was a prospect of improving the path over the railway and so creating the riverside link between the park and the Sainsbury's development.

Other changes in direction resulted from immediate threats of development. One example was at Avebury, where the Trust had since 1943 owned the 4000-year-old Stone Circle, the largest such monument in the world and part of the World Heritage site, but had declined to buy the manor house near by, thinking that its safety could look after itself. This confidence proved misplaced. In 1988-9 the manor was bought by an entrepreneur who built a mock-Tudor market square, a plant centre and other structures within its garden, all without planning permission. At about the same time there were four applications for developments within or adjoining the World Heritage site. A local protest group was formed, the applications were turned down, and in order to remove the eyesores and safeguard the prehistoric landscape for the future the Trust raised a public appeal to buy a semi-derelict café on the Ridgeway (the site of one proposed hotel), farmland lying across the West Kennet Avenue and, with the help of a fortuitous legacy, Avebury Manor itself. The experience showed how, when there was an acute and present danger, the public, English Heritage and the Countryside Commission would come to help; but it also illustrated that without ownership even the most strictly protected areas can suddenly and without warning be at risk.

Threats, I discovered, are not only man-made, nor do they occur only by design. They also arise through acts of nature or of human carelessness. There were three disasters which hit the Trust during my five years. The first was in October 1987 when the Trust suffered the greatest weather damage in its history: 250,000 trees were blown down and many parks and gardens were devastated. London and the southern counties of Hampshire, Sussex, Surrey and Kent suffered the brunt of the storm. The sight of the great beech trees at Slindon snapped off and mutilated as if on the battlefield of the Somme and of the magnificent conifers at Sheffield Park lying flat is not easily forgotten. Not until the roads had been cleared and the most dangerous trees made safe could the foresters and gardeners begin to think of the new opportunities offered by the unwished-for clearings, the propagation of seeds collected from special trees and the planning of new woodlands.

Then in January 1990 violent storms struck again. The area of worst damage was further west and the Trust suffered somewhat less, although it lost another 80,000 trees and the rare shrubs of the Cornish gardens, exempt in 1987, were churned into tangled masses.

Still more upsetting, for it turned upon avoidable human negligence rather than unavoidable elemental force, was the devastation by fire of Uppark in West Sussex. On 30 August 1989 a builder's blowlamp caused a fire which destroyed the upper floors and their contents and badly damaged the state rooms of one of the most untouched and evocative of the Trust's country houses. The lichened surfaces of the exterior walls were miraculously spared, some of the interior plaster-work and decoration on the ground floor survived and almost all the contents of the ground floor and the basement were rescued by the fire brigade and their helpers, and were then carefully packed by the conservators who arrived within hours from all over the country. Next morning, still smouldering, the interior blackened and the floors full of charcoal sludge, the house was dangerously unstable. A local MP called for the remains to be razed to the ground and others suggested that it be left as a ruin or that modern rooms should be constructed within the shell. As soon as I saw how much was left, I had no doubt that it must be restored as nearly as possible to its previous appearance. To demolish or leave it as a ruin would have been unthinkable. The Council shared this view. Today the house once again looks out across the downs, crowned with a new roof, its interior a demonstration that today's plasterers, wood carvers and other craftsmen are as skilled as were their predecessors 250 years ago.

The National Trust with its two million members and land in every county is at once a national institution and a private body, jealous of its independence from government. It cannot be a self-contained island in Britain any more than Britain can be an isolated island in the world. The Trust has dealings with local authorities, government and Parliament, with the European Union and with wider international organizations. In its turn it is subject to pressures from other interests – from the ramblers to open more footpaths, from amenity societies to acquire threatened buildings, from wildlife trusts to protect endangered habitats, from animal-rights groups to ban hunting, from other heritage organizations to take on tasks that they themselves cannot afford or lack the power to carry out.

Although the campaigning role of the Trust's early days has now been supplanted by the CPRE and other bodies, the Trust has continued to use what influence it can wield on general conservation issues. Usually its views are not made public, but in 1987 such a series of measures were announced, affecting conservation in general and the Trust in particular, that the Council decided to speak with as much resonance as it could command.

The first of these measures was a draft circular relaxing planning controls on most agricultural land. In concert with other conservation organizations the Trust opposed the proposals and was successful in securing some improvements, including an unqualified statement that all countryside merited protection 'for its own sake', irrespective of its productive capacity.

The second measure was the privatization of the water authorities. These bodies had owned and protected large areas of outstanding natural landscape for a century and more. The Trust, like the Countryside Commission and other organizations, was alarmed at the possible consequences and lobbied hard to strengthen the safeguards for conservation and public access. The Council's action was strongly supported at the next annual general meeting when, in a strong show of unity between the platform and the floor, members urged that priority should be given to acquiring land disposed of by the new companies. Apart from the transfer of a large area of unspoilt moorland in mid-Wales to a charitable trust, disposals have in fact proceeded slowly, but the Trust has been able to acquire several pockets of dramatic landscape in the Peak District.

Had the properties of the Trust not been protected by the principle of inalienability, the drive to push everything out of public service and into profit-making ownership could have had serious consequences for the Trust. Soon after he became Secretary of State for the Environment, Nicholas Ridley, with his quizzical iconoclasm, asked why we did not sell off some of our country houses and spend the proceeds on repairing the others, thus avoiding any need for government grants. As a former member of the Executive Committee he should have known that this process of cannibalization would not be legal. And he might also have recalled the disastrous example of Heveningham Hall in Suffolk, bought by the government in 1970 and managed by the Trust until it was sold in 1980 and since subjected to most of the vicissitudes to which a great house – and Heveningham is certainly great and exceptionally beautiful – is heir.

In 1990 the announcement of an expanded roads programme threatened large areas of countryside, including thirty of the Trust's own properties. The Trust registered strong objection to an excessively road-based transport policy and to a number of schemes which would affect particular properties, including the Golden Cap estate in Dorset, Cissbury Ring in Sussex and Dunham Massey in Cheshire. The Trust must never be party political, but neither should it be inhibited from speaking out forcefully from time to time on matters of general conservation policy as well as in its own interests.

By the late 1980s it was clear that the Trust should establish regular contacts with the European Commission. Agricultural and environmental policies were now determined to a considerable degree by the Community, and there were opportunities for the Trust to benefit from European funds. In April 1989 Marcus Worsley, Angus Stirling and I paid the first official visit to Brussels on behalf of the Trust. We were accompanied by Leslie McCracken, the international secretary, and by Christopher Audland, who had recently joined the North-West Regional Committee after serving as the Commission's deputy secretary-general and then its director-general for energy and was able to open doors for us. We were pleasantly surprised to discover the interest which the Trust commanded in Brussels. Commissioners and senior officials looked with respect on the Trust as the largest non-government environmental organization in Europe and were impressed by its achievements in the preservation of the coast and the countryside. Our exchange of views on agriculture, training and other matters was followed by a continuing dialogue and by the first grants from the Community to the Trust. The most substantial grants went to the Crom estate in Northern Ireland, the Aberdulais water-power project in South Wales and the Souter Lighthouse in Tyne and Wear, all areas of high unemployment.

Encouraged by this visit, the Trust went ahead with arranging a conference for organizations concerned with the built and natural heritage throughout the whole of Europe. Making possible the attendance of representatives from Central and Eastern Europe was one of the ways in which the Trust has been able to help conservation organizations in other parts of the world. There are now well-established 'national trusts' inspired by the British example in the United States, Canada, Australia, New Zealand and in some islands of the West Indies and other Commonwealth countries. Recent developments have been the founding of the National Trust for Art and Cultural Heritage in India and the amalgamation in Japan of a number of small local trusts into a nationwide organization.

One of the satisfactions of being an officer of the National Trust was that the Trust was firmly in the category of those public service activities at which Britain was pre-eminent. The Latins left much more to the state. The Americans were more fragmented. As an independent organization the National Trust has no rivals for size or comprehensiveness and is, in a minor key, as much a role-model for others as was the British Parliament in the high days of the nineteenth century.

EPILOGUE

THE TRUST was set up to promote the permanent preservation of places of natural beauty or historic interest. This is as urgent a task as it was a hundred years ago, perhaps more so as pressures from people and pollution have intensified. But the way in which the Trust fulfils its purpose has changed. Until quite recently the Trust was judged (and judged itself) by the flow of new acquisitions. Now, having become the largest private landowner in the country, it is judged primarily by the management of the land and buildings it already owns and by the enjoyment it offers to the public.

This does not mean that the Trust should retreat within its present frontiers and say 'no further'. Its benevolent wing has proved far more effective than planning legislation in protecting fine landscapes and monuments. It has a duty to accept gifts of property (if of the right standard and adequately endowed) and to seek funds for new purchases. The fact that purchases are becoming proportionately more important than gifts will allow the Trust to exercise more positive control over where its properties are located (as it has done on the coast for the past thirty years).

Most of the Trust's energy is now directed to maintaining and improving its diverse and widely scattered estates. The standards achieved affect their attraction for the visiting public and can set examples of good practice, as has been done for the conservation of historic houses, the restoration of eroded uplands and the management of heavily visited sites. Now there is an opportunity for the Trust to assume a more ambitious and more experimental role. Could it not, for example, seek solutions to the fundamental question of the revolution in agriculture and its effect on the rural economy:

how farmers can gain a viable living and maintain traditional features in the landscape without polluting the waterways, impoverishing the soil and destroying traditional habitats?

It is the scale of the Trust's practical experience and its exemplary role that gives it an authoritative voice in the environmental movement. Still a crusading organization, but not, except occasionally, a lobbying one, it can and does exert real influence when it intervenes on important planning, agricultural or transport issues or on legislation affecting the conservation of land or historic buildings.

Most people visit the Trust's properties not to study conservation in action but to enjoy a day in the country, or to see a historic house or garden, an ancient monument or an early industrial site. Much has been done to make the properties more welcoming for visitors of all ages and more instructive for those who come to learn. But there is much more still to do. The intention to mark the Trust's centenary with a new initiative to create centres for 'lifelong learning' could give an impetus to developing its resources for educational purposes. The centres will make use of the Trust's wide range of sites: Wicken Fen and other open spaces, Roman remains and ruined abbeys as well as country houses with outstanding collections, unusual historical associations or pioneering estates.

The accusation that the Trust has promoted (and been a beneficiary) of the 'romantic, deferential and fictional approach to the country house'[1] is not without a grain of truth. But this is not to say that country houses should be demolished and their contents sold off, or that they should not open their doors to visitors and become cultural centres. To ignore their unique artistic achievement would be as perverse as to ignore that of the medieval church. The test will be whether the Trust's educational centres can inspire the creative arts and stimulate a deeper concern for the environment rather than encouraging a retreat into a backward-looking nostalgia.

To raise its sights still further ought not to be beyond the capacity of the Trust. It is now an established institution in danger neither from nationalization on the one hand nor from commercialization on the other. Nonetheless, it is a private charity with disposable funds which are small in relation to its responsibility for preserving its properties for posterity; it is dependent on parliamentary support for the right to keep its land inalienable and on government support for the tax reliefs and grants needed for the maintenance of existing properties and the acquisition of new ones. There is no guarantee that either will continue as they are.

A warning shot was fired across the bows of the National Trust for Scotland in 1975 when, after it had resisted plans for building an oil platform on its land, an Act was passed enabling the Secretary of State to acquire land compulsorily and expeditiously for oil-related purposes. The Act has not been used but was an indication that the government would not allow the principle of inalienability to obstruct developments regarded as of overriding national importance. Wherever land is required for public purposes the Trust has to assess the degree of damage to the property concerned and judge how far its objections should be taken. Some accommodation has to be made for small schemes in order not to dissipate the Trust's influence at public enquiries into big schemes or jeopardize the final deterrent of the full parliamentary procedure.

Financial support from government can never be taken for granted in the face of the Treasury's unremitting search for savings. Grants are more vulnerable than tax reliefs and since 1987 their total has declined significantly in real terms. But tax reliefs also are not immune, as recent changes to the regulations applied to covenanted subscriptions have shown. With other sources of revenue stationary or declining, the Trust's financial position is less assured than it has been for some years.

Support from Parliament and government depends ulti-mately on support from the public, and this has continually to be earned. Judged purely in terms of membership, after exploding in the 1980s it has, at least temporarily, reached a plateau. The same is true of visits to properties at which a charge is made. However, the use of its open spaces where access is free continues to rise and so fulfil the aims of its founders.

Increasing size brings practical problems in its wake: the danger of excessive bureaucracy, resentment from local communities at the Trust's dominant position, attacks from special-interest groups, tension between the rights of members and the statutory responsibility of the Council. All these will need sensitive handling and a willingness to embrace change, perhaps in the direction of greater devolution. But the Trust must never be so preoccupied with such day-to-day matters as to lose its vision or its crusading zeal to save 'many a lovely view or old ruin or manor house from destruction and for the everlasting delight of thousands of the people of these islands'.[2]

Officers of the Trust, 1895–1994

PRESIDENTS

Duke of Westminster	1895–1900
Marquess of Dufferin and Ava	1900–1902
Princess Louise, Duchess of Argyll	1902–1944
HM Queen Mary	1944–1953
HM Queen Elizabeth, the Queen Mother	1953–

CHAIRMEN OF THE EXECUTIVE COMMITTEE

Sir Robert Hunter	1895–1913
Earl of Plymouth	1914–1923
John Bailey	1923–1931
Marquess of Zetland	1932–1945
Earl of Crawford and Balcarres	1945–1965
Earl of Antrim	1965–1977
Lord Gibson	1977–1986
Dame Jennifer Jenkins	1986–1990
Lord Chorley	1990–

TREASURERS

Harriot Yorke	1895–1924
Hon. Sidney Peel	1925–1932
Cecil Lubbock	1933–1945
Edward Holland-Martin	1945–1969

CHAIRMEN OF FINANCE COMMITTEE (REPLACING TREASURER)

Mark Norman	1969–1979
Nicholas Baring	1979–1991
Charles Nunneley	1991–

HONORARY SECRETARY

Hardwicke Rawnsley	1895–1920

SECRETARIES/DIRECTOR-GENERALS

Lawrence Chubb	1895–1896
A. M. Poynter	1896–1899
Hugh Blakiston	1899–1901
Nigel Bond	1901–1911
S. H. Hamer	1911–1933
D. M. Matheson	1934–1945
George Mallaby	1945–1946
Vice-Admiral Oliver Bevir	1946–1949
J. F. W. Rathbone	1949–1968
Sir John Winnifrith	1968–1971
Sir Frederick Bishop	1972–1975
Sir Jack Boles	1975–1983
Sir Angus Stirling	1983–

Trust Membership, Acreage and Staff, 1895–1990

Year	Membership	Acreage owned	Acreage* covenanted	Staff† (full time, permanent)
1895	100	5	–	1
1905	500	1,500	–	2
1915	700	6,000	–	2
1925	850	21,000	–	4
1935	4,850	50,000	8,000	5
1945	7,850	112,000	40,000	15
1955	55,658	218,000	41,000	100
1965	157,581	328,500	53,000	450
1975	539,285	383,000	68,000	1,146
1985	1,322,996	533,000	77,000	1,750
1990	2,031,743	569,000	78,000	2,747

*Figures estimated up to 1975
†Figures estimated up to 1965

REFERENCES

CHAPTER ONE
The Founders

1. Charles Dickens, *Pickwick Papers*, chapter 49.
2. Dorothy Hunter, unpublished biography of her father, NT Archives, Acc. 14.
3. *Six Essays on Commons Preservation* (Sampson, Low & Marston, 1867), p. 368, quoted in Graham Murphy, *Founders of the National Trust* (Christopher Helm, 1987).
4. H. D. Rawnsley, 'Robert Hunter: A National Benefactor', *Cornhill Magazine*, February 1913.
5. Hunter, unpublished biography of her father.
6. *The Times*, 7 November 1913.
7. Gillian Darley, *Octavia Hill: A Life* (Constable, 1990), p. 34.
8. Octavia Hill, *Homes of the London Poor* (Macmillan, 1875).
9. Darley, *Octavia Hill*, p. 146.
10. Ibid., p. 310.
11. E. Moberly Bell, *Octavia Hill: A Biography* (Constable, 1942), p. 240.
12. Mary Stocks, *My Commonplace Book* (Peter Davies, 1970), pp. 56–7.
13. Darley, *Octavia Hill*, p. 297.
14. E. F. Rawnsley, *Canon Rawnsley: An Account of His Life* (MacLehose, Jackson, 1923), pp. 12–13.
15. Ibid., p. 26.
16. Ibid., p. 33
17. Ibid., p. 52.
18. *Contemporary Review*, September 1886.
19. Rawnsley, *Canon Rawnsley*, p. 71.
20. Ibid., p. 72.
21. Ibid., p. 78.
22. NT Archives, Acc. 6/1, 25 November 1898.
23. Rawnsley, *Canon Rawnsley*, p. 150.
24. Ibid., p. 200.
25. Ibid., p. 224.
26. Darley, *Octavia Hill*, p. 314.
27. Mark Girouard, *The Victorian Country House* (Yale University Press, 1979), p. 2.
28. *The Times*, 23 December 1899.
29. Rawnsley, *Canon Rawnsley*, p. 26.

30. *SPAB Manifesto* (1877).
31. Philip Henderson, *William Morris: His Life, Work and Friends* (Thames & Hudson, 1967), p. 195.
32. C. Wetherell, *On Restoration* (1857), quoted in James Pope-Hennessy, *Aspects of Provence* (Longmans, 1952).
33. W. Morris, *Athenaeum*, 10 March 1877.
34. W. G. Hoskins and L. Dudley Stamp, *The Common Lands of England and Wales* (Collins, 1963), p. 55.
35. NT Archives, Acc. 6.
36. NT Archives, Acc. 2/1.
37. *The Times*, 17 November 1893.
38. *Daily News*, 17 November 1893.

CHAPTER TWO
Early Years

1. *The Times*, 17 July 1894.
2. Ibid.
3. NT Archives, Acc. 2/4.
4. *Punch*, 21 July 1894.
5. NT Annual Report, 1918–19.
6. NT Archives, Acc. 2/7.
7. H. D. Rawnsley, *A Nation's Heritage* (George Allen & Unwin, 1920), p. 19.
8. Ibid., p. 20.
9. NT Archives, Acc. 6/29.
10. NT Archives, Acc. 6/29(a).
11. NT Archives, Acc. 3/2.
12. NT Archives, Acc. 42/19.
13. *Parliamentary Debates*, House of Lords, vol. XI, 13 February–3 May 1913.
14. NT Annual Report, 1902–3.
15. NT Archives, Acc. 6/21.
16. J. Wake, *Princess Louise: Queen Victoria's Unconventional Daughter* (Collins, 1988), pp. 383–4.
17. NT Annual Report, 1939–40.
18. B. L. Thompson, *Lake District and the National Trust* (Titus Wilson, Kendal, 1946), p. 43.
19. NT Annual Report, 1903–4.
20. Nigel Bond, quoted in W. Creese, *The Search for Environment* (New Haven Press, 1966), p. 223.
21. NT Archives, Acc. 6/49.
22. *Country Life*, 19 September 1904.

23. R. S. T. Chorley, first Baron Chorley, unpublished memoir, NT Archives, Acc. 34.
24. Section 4, National Trust Act 1907.
25. M. Rothschild, *Nathaniel Charles Rothschild, 1877–1923* (Cambridge University Press, 1979).
26. R. Davis, *The English Rothschilds* (Collins, 1983), p. 245.
27. D. Evans, *A History of Nature Conservation in Britain* (Routledge, 1992), pp. 46–9.
28. NT Annual Report, 1909–10.
29. John Gaze, *Figures in a Landscape* (Barrie & Jenkins, 1988), p. 110.
30. Ibid., p. 108.
31. Annual General Meeting, Commons and Footpaths Preservation Society, 7 May 1913.
32. *Spectator*, 15 November 1913.
33. *Westminster Gazette*, 10 November 1914.
34. Robin Fedden, *The Continuing Purpose* (Longmans, 1967), p. 17.
35. NT Archives, Acc. 34.
36. Anne Acland, *A Devon Family* (Phillimore, 1981), p. 150.
37. *The Times*, 22 February 1917.
38. Executive Committee Minutes, December 1919, NT Archives.

CHAPTER THREE
Urban Sprawl

1. NT Archives, 912.
2. Charles Masterman, *England After the War* (Hodder & Stoughton, 1922).
3. Mark Girouard, quoted in David Cannadine, *Pleasures of the Past* (Fontana Press, 1990), p. 99.
4. Stanley Baldwin, *On England* (P. Allan, 1926), pp. 6–7.
5. NT Annual Report, 1920–1.
6. *The Times*, 29 May 1920.
7. *John Bailey 1864–1931: Letters and Diaries* (John Murray, 1935), prefatory note by G. M. Trevelyan, p. 7.
8. Ibid., pp. 229–30.
9. R. S. T. Chorley, unpublished memoir.
10. Ibid.
11. Ibid.
12. Ibid.
13. Executive Committee Minutes, 16 December 1919, NT Archives.
14. R. S. T. Chorley, unpublished memoir.
15. R. C. Norman, unpublished memoir, NT Archives, Acc. 10.
16. NT Annual Report, 1925–6.

17. Ibid.
18. Marquis Curzon and H. A. Tipping, *Tattershall Castle* (Cape, 1929), p. 142.
19. NT Annual Report, 1925–6.
20. *The Times*, 6 February 1924.
21. David Cannadine, *G. M. Trevelyan: A Life in History* (HarperCollins, 1992), p. 12.
22. G. M. Trevelyan, *History of England* (Longmans, 1926), p. 87.
23. G. M. Trevelyan, 'Call and Claims of Natural Beauty', Rickman Godlee Lecture for 1931, in *An Autobiography and Other Essays* (Longmans, 1949), p. 106.
24. G. M. Trevelyan, *Must England's Beauty Perish?* (Faber & Gwyer, 1929).
25. Trevelyan, *An Autobiography and Other Essays*, p. 41.
26. Cannadine, *Trevelyan*. p. 143.
27. Trevelyan, *An Autobiography and Other Essays*, p. 44.
28. Mary Moorman, *George Macaulay Trevelyan* (Hamish Hamilton, 1980), p. 219.
29. Margaret Lane, *The Tale of Beatrix Potter* (Frederick Warne, 1962).
30. Unpublished letters from Mrs Heelis, 1925–39, NT Archives, 157.
31. Ibid.
32. Ibid.
33. Ibid.
34. *Daily Telegraph*, 7 December 1926.
35. Clough Williams-Ellis, *England and the Octopus* (Geoffrey Bles, 1928), p 15.
36. George S. Russell, *National Trust for Scotland: The Formative Years, 1929–1939* (National Trust for Scotland, 1990).
37. Cannadine, *Trevelyan*, p. 156.
38. Ibid., p. 177.
39. NT Annual Report, 1930–1.
40. *NT Magazine*, Spring 1989.
41. Dorothy Hunter, Notes about Ferguson's Gang, NT Archives, Acc. 8.
42. NT Archives, 166.
43. Trevelyan, *Must England's Beauty Perish?*

CHAPTER FOUR
The Country-House Scheme

1. Williams-Ellis, *England and the Octopus*, p. 80.
2. *Country Life*, 25 January 1930.
3. Quoted by Lord Lothian in his speech to the National Trust AGM 1934, NT Archives, Acc. 28/KL/3.

4. R. S. T. Chorley, unpublished memoir.
5. James Lees-Milne, *People and Places* (John Murray, 1992), p. 7.
6. J. R. M. Butler, *Lord Lothian* (Macmillan, 1960), p. 145.
7. Lord Lothian at the National Trust AGM 1934, NT Archives, Acc. 28/KL/3.
8. R. S. T. Chorley, unpublished memoir.
9. Lees-Milne, *People and Places*, p. 4.
10. NT Annual Report, 1935–6.
11. J. M. Keynes, 'Art and the State', in Clough Williams-Ellis (ed.), *Britain and the Beast* (J. M. Dent, 1937), p. 3.
12. Lees-Milne, *People and Places*, p. 8.
13. James Lees-Milne, *Another Self* (Hamish Hamilton, 1970), p. 94.
14. Lees-Milne, *People and Places*, p. 9.
15. Letter from D. M. Matheson to Lord Esher, 14 January 1937, Esher Letters.
16. *Parliamentary Debates*, House of Lords debate on National Trust Act 1939, 30 March 1939.
17. Lees-Milne, *People and Places*, p. 7.
18. *National Trust Guide* (1st edn, National Trust, 1973).
19. James Lees-Milne, *Ancestral Voices* (Chatto & Windus, 1975), pp. 110–12.
20. Speech by Sir Charles Trevelyan, 1934, NT Archives, 40.
21. NT Archives, 40.
22. NT Archives, 878.
23. Ibid.
24. R. S. T. Chorley, unpublished memoir.
25. Lees-Milne, *Ancestral Voices*, p. 282.
26. James Lees-Milne, *Prophesying Peace* (Chatto & Windus, 1977), p. 102.
27. Letter from R. C. Norman to Lord Esher, 8 January 1940, Esher Letters.
28. Letter from Macleod Matheson to James Lees-Milne, NT Archives 878.
29. NT Archives, 36.
30. Letter from Earl Spencer to Lord Esher, 12 August 1942, Esher Letters.
31. Lees-Milne, *Ancestral Voices*, p. 27.
32. Lees-Milne, *Prophesying Peace*, p. 140.
33. Ibid., p. 92.
34. *Daily Mirror*, 1944, undated clipping, NT Archives, 36.
35. NT Annual Report, 1944–5.
36. NT Archives, 2223.
37. NT Archives, 36.
38. Lees-Milne, *Prophesying Peace*, p. 204.
39. Ibid., p. 197.

●

CHAPTER FIVE
Subtopia and the Second World War

1. A. G. Street, 'The Countryman's View', in Williams-Ellis (ed.), *Britain and the Beast*, pp. 122–32.
2. C. F. Carr, *The Complete Hiker and Camper* (Pitman, 1932), p. 119.
3. Marion Shoard, *This Land is Our Land* (Paladin, 1987), pp. 112–14.
4. NT Sound Archives.
5. *Daily Herald*, 16 June 1937.
6. Darley, *Octavia Hill*, p. 304.
7. *Journal of the Royal Society of Arts*, 6 May 1926.
8. NT Archives, 637.
9. NT Annual Report, 1935–6.
10. Christopher Gibbs, 'Personal Memories of Early Days with the Trust', NT Archives, Acc. 33.
11. NT Archives, 1081.
12. NT Annual Report, 1935–6.
13. NT Annual Report, 1939–40.
14. NT Annual Report, 1941–2.
15. *The Times*, 1 February 1944.
16. *The Times*, 3 February 1944.
17. *The Times*, 4 February 1944.
18. Lees-Milne, *Prophesying Peace*, p. 62.
19. NT Sound Archives.
20. NT Archives, 319.
21. NT Archives, 2223.

CHAPTER SIX
Peace and Socialist Utopia

1. Zetland Papers.
2. R. S. T. Chorley, unpublished memoir.
3. Crawford Papers.
4. Ibid.
5. R. S. T. Chorley, unpublished memoir.
6. Crawford Papers.
7. Gaze, *Figures in a Landscape*, p. 36.
8. Crawford Papers.
9. Ibid.
10. *The Times*, 9 October 1963.
11. 13 July 1946, Esher Letters.
12. James Lees-Milne, *Caves of Ice* (Chatto & Windus, 1975), p. 118.

13. Crawford Papers.
14. Ibid.
15. Ibid.
16. Ibid.
17. 12 June 1945, Esher Letters.
18. Crawford Papers.
19. Lees-Milne, *Caves of Ice*, p. 47.
20. Ibid., p. 59.
21. Ibid., p. 64.
22. 27 June 1948, Esher Letters.
23. Gaze, *Figures in a Landscape*, p. 186.
24. Rosse Papers.
25. R. C. Norman, unpublished memoir.
26. NT Archives, 2223.
27. Lees-Milne, *Caves of Ice*, p. 102.
28 Lionel Robbins, quoted by Hugh Dalton, *High Tide and After* (Frederick Muller, 1962), p. 119.
29. Ben Pimlott, *Dalton* (Cape, 1985), p. 455.
30. *Parliamentary Debates*, House of Commons, vol. 421, 9 April 1946.
31. Crawford Papers.
32. Introduction to Williams-Ellis (ed.), *Britain and the Beast*, p. viii.
33. James Lees-Milne, *Midway on the Waves* (Faber & Faber, 1985), p. 35.
34. NT Annual Report, 1947–8.
35. R. S. T. Chorley, unpublished memoir.
36. John Dower, *National Parks in England and Wales* (Ministry of Town and Country Planning, 1945).
37. NT Archives, 319.
38. Letter from J. F. W. Rathbone to Lord Esher, NT Archives, 319.
39. Dower, *National Parks in England and Wales*.
40. NT Archives, 319.
41. Lees-Milne, *Caves of Ice*, p. 6.
42. 19 January 1949, Esher Letters.
43. Merlin Waterson (ed.), *The Country House Remembered* (Routledge & Kegan Paul, 1985), p. 20.

CHAPTER SEVEN
White Elephants

1. Lees-Milne, *People and Places*, p. 177.
2. Ibid., p. 151.
3. John Wyndham, *Wyndham and Children First* (Macmillan, 1968).
4. Ibid.

5. Crawford Papers.
6. Lees-Milne, *Ancestral Voices*, p. 273.
7. NT Archives, 1409.
8. Lees-Milne, *Ancestral Voices*, pp. 171–2.
9. NT Archives, Historic Houses 36/3.
10. Crawford Papers.
11. Lord Ronald Gower, *My Reminiscences* (Kegan Paul, Trench, 1883), pp. 476–7.
12. NT Archives, Chartwell File.
13. NT Annual Report, 1951–2.
14. 29 January 1949, Esher Letters.
15. 1 February 1949, Esher Letters.
16. NT Archives, Historic Houses 36/3.
17. Ibid.
18. Ibid.
19. NT Archives, 1760.
20. Ibid.
21. NT Archives, 1412.
22. NT Annual Report, 1947–8.
23. NT Annual Report, 1949–50.
24. 25 April 1948, Esher Letters.

CHAPTER EIGHT
Western Approaches

1. Lees-Milne, *People and Places*, p. 180.
2. 13 July 1955, Esher Letters.
3. Valedictory address by Sir Ralph Verney to the Regional Committee, 16 July 1993.
4. 9 November 1956, Rosse Papers.
5. *Field*, 22 August 1957.
6. Robert Latham, National Trust Sound Archives, no. 232.
7. *The Times*, 2 July 1979.
8. Roy Strong (ed.), *Destruction of the Country House, 1875–1974* (Thames & Hudson, 1974).
9. 1958, Esher Letters.
10. Strong (ed.), *Destruction of the Country House*.
11. Evelyn Waugh, Foreword to *Brideshead Revisited* (rev. edn, 1959).
12. 28 July 1957, Esher Letters.
13. Mrs James de Rothschild, *The Rothschilds at Waddesdon Manor* (Collins, 1979), p. 126.
14. 15 February 1959, Crawford Papers.

15. 10 August 1956, Crawford Papers.
16. Clough Williams-Ellis, *In Trust for the Nation* (Paul Elek, 1947).
17. NT Archives, File 1957.
18. *NT Newsletter*, Spring 1962.
19. Ibid.
20. NT Archives, 2044.
21. Gaze, *Figures in a Landscape*, p. 273.
22. Fedden, *The Continuing Purpose*, p. 52,
23. 7 December 1949, Rosse Papers.
24. 22 June 1959, Antrim Papers.
25. 18 September 1959, Antrim Papers.
26. 21 September 1959, Antrim Papers.
27. 14 October 1964, Antrim Papers.

CHAPTER NINE
Gardens and Quirkeries

1. *Estates Gazette*, July 1952.
2. NT Archives, 2322.
3. NT Archives, 2261.
4. 15 December 1952, Esher Letters.
5. NT Annual Report, 1951–2.
6. Antrim Papers.
7. *Sunday Express*, 6 September 1959.
8. 18 September 1959, Antrim Papers.
9. NT Archives, 2223.
10. NT Archives, 2223.
11. 22 July 1960, Rosse Papers.
12. NT Archives, 2223.
13. Ibid.
14. Ibid.
15. 16 April 1963, Crawford Papers.
16. 6 December 1963, Crawford Papers.
17. 13 December 1963, Rosse Papers.
18. Mark Norman, unpublished memoir.
19. Elizabeth Battrick, *Guardian of the Lakes* (Westmorland Gazette, 1987), p. 22.
20. Ibid., p. 18.
21. Anne Scott-James, *Sissinghurst: The Making of a Garden* (Joseph, 1974), p. 16.
22. Ibid., p. 122.
23. Nigel Nicolson (ed.), *Harold Nicolson, Diaries and Letters, 1945–62* (Collins, 1968), p. 24.

24. NT Archives, 1802/1.
25. Nigel Nicolson, *NT Magazine*, Summer 1970.
26. 17 November 1961, Rosse Papers.
27. NT Archives, 1802/1.
28. Ibid.
29. Ibid.
30. Ibid.
31. *Daily Mirror*, 17 August 1971.
32. *NT Magazine*, Spring 1968.
33. NT Archives, 472.
34. NT Annual Report, 1964–5.
35. 3 May 1965, Crawford Papers.
36. 8 March 1965, Crawford Papers.
37. 1 May 1965, Crawford Papers.
38. Mark Norman, unpublished memoir.
39. 19 July 1965, Crawford Papers.

CHAPTER TEN
Admiral Neptune and Commander Rawnsley

1. Christopher Gibbs, unpublished memoir, NT Archives, Acc. 33.
2. NT Archives, EN/C2/1.
3. Ibid.
4. Speech by Prince Philip, NT Annual Report, 1964–5.
5. NT Archives, EN/C4/8.
6. 19 December 1963, Crawford Papers.
7. NT Archives, EN/C4/8.
8. NT Archives, EN/C4/9.
9. Ibid.
10. NT Archives, EN/C5/3.
11. NT Archives, EN/C4/8.
12. NT Archives, EN/F9.
13. Crawford Papers.
14. NT Archives, EN/F9.
15. *Observer*, 6 November 1966.
16. *Guardian*, 7 November 1966.
17. NT Archives, AGM 1966.
18. Ibid.
19. *Guardian*, 14 November 1966.
20. NT Archives, EN/F9.
21. NT Archives, AGM 1966.
22. NT Archives, EN/F9.

23. Ibid.
24. NT Archives, EN/F8.
25. NT Archives, EN/F9.
26. *Guardian*, 19 January 1967.
27. NT Archives, EN/F8.
28. NT Archives, EN/F8.
29. Minutes of EGM, 11 February 1967, NT Archives.
30. NT Archives, EN/F2/2.
31. Letter from Lord Antrim to E. R. Cochrane, 3 March 1967 (copy in Grigg Papers).
32. *Observer*, 12 February 1967.
33. Letter to Len Clark, 26 June 1967.
34. NT Archives, EN/C4/9.

CHAPTER ELEVEN
Trial by Benson

1. Lord Benson, *Accounting for Life* (Kogan Page, 1989), p. 147.
2. Gaze, *Figures in a Landscape*, p. 228.
3. Letter from E. R. Cochrane to John Grigg, Grigg Papers.
4. Benson, *Accounting for Life*, p. 149.
5. Benson Report, para. 133 (1968).
6. NT Archives.
7. Rosse Papers.
8. 12 April 1978, Rosse Papers.
9. NT Archives, Acc. 11/14.
10. NT Archives, Acc. 11/18.
11. Ibid.
12. Mark Norman, unpublished memoir.
13. *A Modern Major General: The Memoirs of Julian Gascoigne, 1903–1990* (private pub.), p. 109.
14. Lord Oliver of Aylmerton, *Report on the Workings of the Constitution* (National Trust, 1993), para. 114.
15. 21 November 1967, Rosse Papers.
16. *The Times*, 8 January 1993.
17. *NT Magazine*, Spring 1971.
18. Gaze, *Figures in a Landscape*, p. 244.
19. NT Archives, 2223.
20. NT Annual Report, 1975.
21. Trinick Papers.
22. 16 August 1974, Rosse Papers.
23. 4 September 1971, Rosse Papers.
24. 26 February 1977, Rosse Papers.

CHAPTER TWELVE
The New Professionals

1. Mark Norman, unpublished memoir.
2. Interview with Edward Fawcett, November 1993.
3. Letter from John Hodgson to Jennifer Jenkins, 21 February 1994.

CHAPTER THIRTEEN
Decorating Historic Houses

1. *Country Life*, 15 July 1971.
2. John Cornforth, *Country Life*, 21 February, 1985.
3. The Duchess of Devonshire, *The House: A Portrait of Chatsworth* (Macmillan, 1982), pp. 127–8.
4. NT Archives, Decorating 36/15.
5. Ibid.
6. NT Archives, 2473.
7. Marlin Waterson, *The Servants' Hall* (Routledge & Kegan Paul, 1985)

CHAPTER FOURTEEN
Fora, Fauna and Flora

1. White Paper on Conservation of Nature in England and Wales (Ministry of Town and Country Planning, 1947).
2. National Trust Act 1907.
3. Sir Fraser Darling, BBC Reith Lecture, 1969, quoted in *The National Trust for Scotland Guide* (rev. edn, Jonathan Cape, 1976), p. 307.
4. Oliver Rackham, *The Last Forest* (J. M. Dent, 1989), p. 1.
5. Ibid., p. 152.
6. John Cripwell, *NT Magazine*, Autumn 1982.
7. *Sunday Times*, 10 October 1982.
8. Letter from Audrey Urry to Jennifer Jenkins, 18 December 1993.
9. Letter from Lord Beaumont to Jennifer Jenkins, 23 December 1993.
10. NT Archives, 2093.
11. Ibid.
12. Gervase Jackson-Stops (ed.), *Treasure Houses of Britain* (Yale University Press, 1984), pp. 302–3.

CHAPTER FIFTEEN
A Personal Perspective

1. Keith Thomas, *Man and the Natural World* (Penguin, 1983).
2. Lord Oliver of Aylmerton, *Report on the Workings of the Constitution*, para. 47.

EPILOGUE

1. David Cannadine, *Aspects of Aristocracy* (Yale University Press, 1994).
2. Octavia Hill speaking at the meeting of the Provisional Council, 16 July, 1894 (*The Times*, 17 July 1894).

INDEX

———————————— • ————————————

Abbeydale Steel Works, Sheffield, 193
Aberconway, Charles McLaren, 3rd
    Baron, 159
Aberconway, Henry Duncan McLaren,
    2nd Baron, 157, 159, 185–6, 202
Abercrombie, Patrick, 61–2
Aberdulais water power project, South
    Wales, 330
Abergavenny, Gwent *see* Sugar Loaf
Abergwesyn Commons, Powys, 295, 297
Access to Mountains Act (1939), 105, 123
Acland, Cuthbert H. D. ('Cubby'):
    activities in Lake District, 133, 189;
    meets Silkin, 141; accepts hill farming
    grants, 161; retires, 204; and public
    amenities, 205; authority, 221; and
    Rawnsley's EGM, 235; relations with
    NT committees, 252; and information
    centres, 261
Acland, Francis, 41, 43
Acland, Sir Richard, 41, 116–17, 301
Acland, Sir Thomas Dyke, 41, 45
Acorn Camps, 265
Adam, Robert, 102, 150, 271
Addison, Christopher, First Viscount, 62
Advisory Committee on Natural History
    (NT), 287
Agricultural Holdings Bill (1984), 292
agriculture: inter-war depression, 44;
    modern practices, 286
Agriculture, Board of, 37, 42
Aintree: Grand National racecourse, 305
A La Ronde, Devon, 308
Albeck, Pat, 262
Alfriston, Sussex: clergy house, 26–7, 35,
    304
Alice, Princess, Grand Duchess of Hesse-
    Darmstadt, 32
Allan Bank, Cumbria, 13
Allinson, Neil, 318–19
Althorp, Northamptonshire, 97
Ancient Monuments Acts: (1900), 29;
    (1913), 30; (1931), 69
Ancient Monuments Society, 307
Anglesey Abbey, near Cambridge, 206
Anglesey, George Charles Henry Victor

Paget, 7th Marquess of, 166
animals: rare breeds, 277, 296
Annual Handbook (NT), 264, 266
Antoninus, Wall of (Scotland), 28
Antrim, Randal John Somerled
    McDonnell, 13th Earl of: serves on
    Executive Committee, 128;
    Chairmanship of Northern Ireland
    Committee, 182–7; on Beaverbrook
    criticisms, 196; and Rathbone, 203;
    and conflict with Bath City Council,
    215; succeeds Crawford as chairman
    of the National Trust, 217–18; and
    Enterprise Neptune, 223–4; disputes
    with Conrad Rawnsley, 224, 232, 234,
    239; gives notice to Conrad Rawnsley,
    228; at 1967 EGM, 236–7; and Benson's
    Committee of Enquiry, 238, 240, 243;
    praises Winnifrith, 250; appoints Boles
    director-general, 256; death, 256–7;
    criticizes Fenton House decoration,
    273
Antrobus, Sir Edmund, 29
Apsley House, London, 102, 155
archaeological advisers, 293–4
Ardress House, County Armagh, 212
Arkell, John, 131; chairs committee of
    investigation, 302–3, 310
Arlington Court, Devon, 38–9, 176
Armitage, Joseph, 115
Armstrong, Sir William, 1st Baron
    Armstrong of Cragside, 277–8
Arnold, Matthew, 10
Arts Panel, 270
Ascott, Buckinghamshire, 38
Ashbee, Charles, 32
Ashdown, Oxfordshire, 167–8
Ashridge estate, Hertfordshire, 55–6, 313
Ashton, Sir Leigh, 149
Ashton Wold, Northamptonshire, 37
Asquith, Herbert Henry, 1st Earl of
    Oxford and Asquith, 56
Association for the Protection (*earlier*
    Preservation) of Rural Scotland
    (APRS), 63–4
Astor, Nancy, Viscountess, 14, 98

Astor, Waldorf, 2nd Viscount, 14, 98

Astor, William Waldorf, 1st Viscount, 14

Astor, William Waldorf, 3rd Viscount, 196

Athenaeum (journal), 17

Atholl, George Iain Murry, 10th Duke of, 313

Atholl, John George Stewart-Murray, 8th Duke of, 65

Attingham, Shropshire, 177, 180, 188

Attlee, Clement, 1st Earl, 47, 63, 122, 155

Audland, (Sir) Christopher, 329

Avebury, Wiltshire, 118, 325–6; Stone Circle, 293

Aynhoe House, Oxfordshire, 251

Badbury Rings, Dorset, 296

Baden-Powell, General Robert, 1st Baron, 189

Bailey, John: praises Rawnsley, 13; chairmanship of NT Executive Committee, 41, 46, 48–50, 96; on choice of properties, 42; on Chadwich Manor, 54; and Trevelyan, 56–7; death, 72, 76

Bailey, Sarah (*née* Lyttelton), 46–7

Baines, Sir Frank, 110

Baldwin, Stanley (*later* 1st Earl), 44–5, 52, 55–6, 61, 69, 72, 109–10

Ballymoyer House, County Armagh, 112

Bangor, Agnes Elizabeth, Viscountess, 184

Bankes, Ralph, 296

Bannerman, Alastair, 262

Bannockburn, Battle of, 65

Barber, Anthony, Baron, 251

Baring, Nicholas, 312–14

Barlow Report on the Distribution of the Industrial Population, 138

Barras Head, Cornwall, 27

Barrington Court, Somerset, 35–6, 162

Bateman's Sussex, 212

Bath Assembly Rooms, 72, 189, 266

Bath City Council, 215–16

Bath, Henry Frederick Thynne, 6th Marquess of, 255

Bath, Thomas Henry Thynne, 5th Marquess of, 85, 174, 196

Battle Abbey, Sussex, 305

Bearsted, Walter Horace Samuel, 2nd Viscount, 135

Beaumont, Michael, 260

Beaumont, Timothy, Baron, 300

Beaverbrook, William Maxwell Aitken, 1st Baron: press attacks NT, 196–7

Bedford, John Robert Russell, 13th Duke of, 173

Beeton, David, 313

Bellotto, Bernardo, 280

Belmore, Armar Lowry-Corry, 1st Earl of, 183

Belmore, Galbraith Armar Lowry-Corry, 7th Earl of, 183

Belton House, Lincolnshire, 281–2, 297

Ben Lawers, Scotland, 288

Beningbrough Hall, Yorkshire, 274

Benson, Sir Henry (*later* Baron): Committee and Report on NT (1967–8), 238, 240–53, 259, 304

Berkhamsted Common, Hertfordshire, 20, 55

Berwick, Edith Teresa, Lady, 180

Bethell, Nefertiti, 148

Betjeman, (Sir) John, 172, 182, 187, 237, 277

Bevir, Vice-Admiral Oliver, 131–2, 159

Biggs, Nicholas, 313

Binns, House of, 66

Birmingham City Council, 226

Birr Castle, County Offaly, Ireland, 158, 170–1

Bishop, Sir Frederick, 250–2, 254–6, 258

Blaise Hamlet, Somerset, 118

Blakeney Point, Norfolk, 36, 287

Blenheim Palace, Oxfordshire, 83, 196

Blickling Hall, Norfolk, 78, 98, 157; estate farms, 161; wartime damage, 162; early books, 177; workshops, 270

Blomfield, Ivor, 285, 304, 308, 313

Blunt, Anthony, 135, 269

Bodiam Castle, Sussex, 52

Bodnant, Gwynedd, 157, 159

Boles, Sir John Dennis: as secretary of Benson Committee, 241, 252; succeeds Bishop as secretary of NT, 255–6; recorganizes membership department, 260; and Stirling, 297; and Bradenham disputes, 298, 300, 302–3; demand for resignation, 300

Bolitho, Lieutenant-Colonel Sir Edward, 207

Bollom, Ian, 308–9

Bond, Nigel: on NT founders, 13; on Barrington Court, 36; proposes

general endowment fund, 50; on
    Country House Committee, 83;
    retires, 95
Bonham-Carter, Sir Edgar, 83
Bonham-Christie, John, 190
Bonham-Christie, Mary, 189
Bonner-Pink, Captain Ralph, 249
Borrans Field, Ambleside, Cumbria, 33
Boston Metropolitan Park Commission
    (USA), 22
Boswell, James, 107
botanical advisers, 285
Bowen, Roderick, 181
Bowes-Lyon, Sir David, 158, 209
Box Hill, Surrey, 4, 50, 268
Bradenham estate, Buckinghamshire,
    189, 298, 313; dispute over, 298–301,
    317
Brandelhow Park, Derwentwater,
    Cumbria, 31–3
Bransdale, Yorkshire, 265
Brantwood, Cumbria, 13
Brett, Alfred, 113
Brett, (Sir) Charles, 186
Bridges, (Sir) Edward, 1st Baron, 136,
    156, 237
British Waterways Board, 193
British Workers' Sports Federation, 106
Broadclyst (village), Devon, 118
Brooke, Lord, (later 8th Earl of Warwick),
    174
Brookeborough, Basil Stanlake Brooke,
    1st Viscount, 187
Brotherton, Helen, 190
Brown, John & Co., 157, 159
Brown, Lancelot ('Capability'), 146, 206,
    216
Brownlow, Edward John Peregrine Cust,
    7th Baron, 281–2
Brownsea Island, Dorset, 189–90
Bruce's Stone, Galloway, 65
Bryce, James, Viscount, 21
Buchan, John (1st Baron Tweedsmuir), 72
Budlake (village), Devon, 118
Bury, Lady Mairi, 185–6
Buscot estate, Oxfordshire, 188
Bute, John Crichton-Stuart, 4th Marquess
    of, 66
Butler, Montagu, 13, 24
Butler, Richard Austen, Baron, 137, 165
Buxton, Anthony, 290–1
Buxton, Edward North, 290

Buxton, S. F., 163

Cadbury, Edward and George, 54–5
Cadbury, Michael, 248
Caernarfon: Roman fort, 114
Calke Abbey, Derbyshire, 281–3, 297, 308
Callaghan, James (later Baron), 223
Campaign for Nuclear Disarmament, 299
Camrose, William Ewert Berry, 1st
    Viscount, 151
canals, 190–3
Canna (island), 67
Cannadine, David: The Decline and Fall of
    the British Aristocracy, 75
Canons Ashby, Northamptonshire, 279,
    281–2, 308
caravans, 205, 220, 227
Carlisle, John Wareing Bardsley, Bishop
    of, 10, 13
Carlisle, George James Howard, 4th Earl
    of, 24
Carson, Rachel: Silent Spring, 292
Castle Coole, County Fermanagh, 183
Castle Crag, Borrowdale, Cumbria, 33
Castle Howard, Yorkshire, 135
Castle Ward, County Down, 184
Cerne Giant, Dorset, 51
Chadwich Manor, near Birmingham, 54
Charity Organization Society (COS), 15,
    47
Charlecote, Warwickshire, 162, 167, 177,
    193
Charles II, King, 206
Charlotte, Princess, 196
Charteris of Amisfield, Martin, Baron,
    279, 280
Chartwell, Kent, 150–1
Chase, Ellen, 22
Chastleton House, Oxfordshire, 284, 308
Chatsworth, Derbyshire, 83, 147; opened
    to visitors, 173
Cheddar Cliffs, Somerset, 287
Chedworth Roman Villa,
    Gloucestershire, 51
Cheltenham Annual General Meeting
    (1966), 229, 232
Cheshire County Council, 195
Chesters estate, Northumberland, 57
Chichester, Rosalie, 38
children: NT attitude to, 266–7
Chippendale, Thomas, 283
Chorley, Professor Robert Samuel

Theodore, 1st Baron: serves on Council, 47, 61; on NT management, 48; on Zetland, 77; on Matheson, 81; on aristocratic composition of NT officers, 96; and postwar Labour government, 123; on Crawford, 125–6; appointed to NT committees, 128; seeks Dalton's advice on NT function, 139; and establishment of National Parks, 140–1; on Rathbone, 203; approves Antrim as chairman, 218; speaks at 1967 EGM, 236–7

Chorley, Roger Richard Edward, 2nd Baron: becomes chairman of NT, 47, 245; joins NT Finance Committee, 245; endowment formula, 297; on committee to review staff-committee relations, 304; refers voting demands to Lord Oliver, 318

Chubb, Lawrence, 34, 63

Church of England: and SPAB, 17

Churchill, (Sir) Winston S., 75, 150–1, 155

Cissbury Ring, Sussex, 51, 329

Citrine, Sir Walter, 1st Baron, 123

Clandon Park, Surrey, 196, 271–2

Clanwilliam, John Charles Edmund Carson Meade, 6th Earl of, 187, 247

Claremont Landscape Garden, Surrey, 137, 196

Clark, Leonard: and Rawnsley dispute, 235, 237; serves on Council, 235; serves on Benson Committee, 238, 241, 246–7; on Council practice, 243; on committee to report on staff-committee relations, 304

Claydon, Buckinghamshire, 168, 271

Clifton College Mission, Bristol, 9

Cliveden, Buckinghamshire, 14, 97–8, 157, 162, 196, 206, 211, 271

Clumber Park, Nottinghamshire, 118

coastal (seashore) protection and acquisition , 66–7, 108–9, 199, 319–20; and Enterprise Neptune appeal, 219–23, 238–9

Cobbett, William, 19

Cochrane, Raymond, 232, 238, 241

Coggeshall, Essex, 71

Colefax, Sibyl, Lady, 271

Coleshill estate, Oxfordshire, 188

Colliers Wood, London, 49

Colquhoun of Luss, Sir Iain, 65

Committee for Wales, 113, 181

common land: nature of, 18–20

Common Wealth (political party), 116–17

Commons Preservation Society (CPS), 3, 5, 15, 18, 20–1, 23–4, 27, 55; merges with National Footpaths Preservation Society, 29

Complete Hiker and Rambler, The, 105

conifer plantations, 67–9

Coniston Lake, Cumbria, 205

Conservative party: 1951 government, 155, 165; policy on taxation and grants for country houses, 155; under Thatcher, 253

Constable, John, 118

Constable, W. G., 76

Consumers' Association, 199

Conwy, Gwynedd, 114

Conwy Suspension Bridge, Gwynedd, 193–4

Conway Valley, Gwynedd, 178

Cook, Ernest, 72, 189

Cooper, Sir Francis D'Arcy, 241n

Corfe Castle, Dorset, 296

Corn Production Act (1921), 44

Cornforth, John, 273, 274, 277–8, 311

Cornwall: Trinick's service in, 189, 198, 220–1

Cornwall, Cape (near Land's End), 320

Cotehele, Cornwall, 83, 137, 154

cottages: acquisition and maintenance of, 110

Council (NT): composition, 63, 95, 313–14; election to, 198; status at 1967 EGM, 236–7; sets up 1967 Committee of Enquiry, 237–8; role and responsibilities confirmed by Benson Committee 242–3, 311; and Bradenham dispute, 299, 301–2; meetings, 311, 313

Council for the Preservation of Ancient Bristol, 118

Council for the Protection of Rural England (CPRE): formed (1926), 61; opposes conifer plantations, 67–8; campaigning, 328

Counter Attack (plan), 188

country houses see landed estates

Country Houses Committee (NT), 83

Country Life (magazine): on Barrington Court, 35; on threat to country houses, 74; on Fowler, 273

Country-House Scheme: property

transfers under, 66, 89–95, 97–101, 105; initiated and legalized, 85–7; and family occupancy, 86, 89, 102; and maintenance endowments, 86–7; and house contents and furniture, 87
countryside: threats to, 54, 61–2, 104–5, 188, 308, popular access to, 105–7
Countryside Commission, 210, 239, 288, 295, 326
County Naturalists' Trusts, 288–9
covenants, restrictive, 109
Covington, Stenton, 95
Cragside, Northumberland, 268, 277–8, 308
Craven, Cornelia, Countess of, 168
Crawford, David Alexander Edward Lindsay, 27th Earl of, 61, 65, 88
Crawford, David Robert Alexander Lindsay, 28th Earl of: and National Trust for Scotland, 65; qualities, 77; chairmanship, 124–8; and appointment of officers, 128–9; praises Holland-Martin, 128; on Mallaby's departure, 131; on Rathbone, 132; and Land Fund, 136–7; and Petworth House pictures, 146; and Ham House, 149; and government grants, 152–3; and Gowers Committee recommendations, 154; and government policy, 155; 1948 tour of inspection, 162–3; and Esher's methods, 170; visits Woburn, 173–4; and Waddesdon, 174–6; on maintaining archives and books, 176; John Smith appeals to on canals, 191–2; lacks interest in administration, 198; and social changes, 199; and John Smith's reforms, 200–2; appeals for help for gardens, 210; retires, 217–18; and Enterprise Neptune, 223–5; and Rathbone's retirement, 249
Crawshay, Lieutenant-Colonel W. R., 226–7
Creech Jones, Arthur see Jones, A. C.
Cripps, Sir Stafford, 137, 151
Cripwell, John, 296
Crockham Hill, Kent, 34
Crom estate, County Fermanagh, 320, 330
Crookenden, Spencer, 312
Crosthwaite, Keswick, Cumbria, 10, 12

Culloden, 65
Culross Palace, Scotland, 66
Culross, Royal Burgh of (Scotland), 67
Curzon of Kedleston, George Nathaniel, Marquess, 30, 52–3, 72, 87
Cushendun (village), County Antrim, 187
Cwmdu, Dyfed, 323

Daily Herald, 108
Daily Mirror, 100, 214
Daily News, 23
Daily Telegraph, 61, 196, 226, 234
Dalness, Scotland, 66
Dalton, Hugh, Baron, 30, 123, 135–7, 139–43, 156
Dalton, Ruth, Lady, 126, 129, 141, 225
Darling, Sir Fraser, 288
Dashwood, Sir John, 96, 110
Davis, Warren, 261
De La Warr, Herbrand Edward Dundonald Brassey Sackville, 9th Earl, 96, 128, 134, 163, 202
de Stein, Sir Edward, 101
death duties see taxation
deer hunting, 317–18
Defence, Ministry of: and Bradenham dispute, 298–9, 302
Demeure Historique, La (French organization), 81
Denyer, Susan, 315
Derwent Estate, Derbyshire, 189
Destailleur, Gabriel Hippolyte, 174
Devil's Punch Bowl, Surrey, 4, 34
Devonshire, Andrew Cavendish, 11th Duke of, 147, 169
Devonshire, Deborah, Duchess of, 273
Devonshire, Evelyn, Duchess of, 169
Dimbleby, Richard, 225
Dinas Oleu, Gwynedd: acquired as first NT property, 25–6
Disraeli, Benjamin (Earl of Beaconsfield), 150
Disraeli Society, 150
Dolaucothi estate, near Lampeter, 114
Dolgellau, Gwynedd, 113
Dolmelynllyn, Gwynedd, 113
Donnelly, Desmond, 142
Dorset Naturalists' Trust, 190
Douglas-Pennant, Lady Janet, 178
Dovedale, Derbyshire, 105, 107–8
Dover: White Cliffs, 305
Dower, John, 139–40, 142, 286

Drovers Estate, Sussex, 241n
'Druids Circle', Castlerigg, Cumbria, 33
Drury, Martin, 270, 280, 321
Dryden, family, 279, 281
Dryden, John, 281
Duddon Valley, 58
Dufferin and Ava, Frederick Temple-
    Hamilton-Blackwood, 1st Marquess
    of, 24, 32
Duggan, George, 112
Dunham Massey, Cheshire, 329
Dunkery Beacon, Somerset, 118
Dunstanburgh Castle, Northumberland,
    320
Durham County: coastal acquisition, 320
Dyrham, Avon, 212

East Anglia: regional organization, 198
education: NT's commitment to, 267–8
Edward VII, King, 33
Egremont, 1st Baron see Wyndham, John
Elizabeth, Queen Mother, 217
Ely Cathedral, 17
English, David, 249
English Heritage: incorporates Historic
    Buildings Council, 166; guardianship
    of monastic ruins, 295; and Avebury
    Manor, 326
Enniskillen, Earls of, 184
entailed estates, 88, 97
Epping Forest, Essex, 4, 20
Erddig, Clwyd, 275–6
Esher, Lionel Brett, 4th Viscount, 298, 300
Esher, Oliver Brett, 3rd Viscount: and
    country house crisis, 83, 88–9; chairs
    Finance and General Purposes
    Committee, 96, 127; on shortage of
    country house acquisitions, 97;
    wartime activities for NT, 97, 99; on
    minimum endowments, 98; on Lacock
    village, 111; donates Watlington Hill
    to NT, 118; and Crawford's
    chairmanship, 127; and Holland-
    Martin, 128; qualities, 129; favours
    donor families, 130; on Mallaby's
    departure, 130; approves area
    representatives, 134; and Land Fund,
    137; and NT's inalienable land,
    139–40; and National Parks, 140–1; on
    future of NT, 142; on quality of
    acquisitions, 145; on management of
    unoccupied houses, 152; and

government grants to NT, 155;
    Crawford reports on state of
    properties to, 162; and Fedden's work
    on Ashdown, 168; acquires houses for
    NT, 170; dislikes Waddesdon, 175;
    Knollys complains to on financial
    restrictions, 194; lacks interest in
    administration, 198; Smith succeeds as
    committee chairman, 199–200
Estates Committee (NT), 50, 128, 134,
    219, 244
Estates Gazette, 184, 188
Europa Nostra, 306
European Community, 253, 306, 329–30
Euston, Hugh FitzRoy, Earl of see
    Grafton, 11th Duke of
Evelyn, John: Deptford house, 20
Eversley, Baron see Shaw-Lefevre, George
    John
Ewbank, Sir Robert, 141
Executive Committee: composition, 95,
    129; on management of unoccupied
    houses, 152; accepts Hidcote, 159;
    elects Council members, 198; and seal
    culling, 214; role and responsibilities,
    243–4, 246; meetings, 311–13
Exmoor, 41; red deer, 317
Extraordinary General Meetings:
    Rawnsley calls (Westminster, 1967),
    231, 233, 235–7, 302; Bradenham
    dispute (Wembley, 1982), 300–2, 317

Fair Isle, 67
Fairhaven, Urban H. R. Broughton, 1st
    Baron, 206
Fairlight Cove, Sussex, 220
farm properties: maintenance and
    improvements to, 161, 163, 204–5, 253;
    and wildlife protection, 292–3
Farne Islands, Northumberland, 50, 55,
    287–8; seal culling, 199, 213–14, 217
Fawcett, Edward: as director of public
    relations, 258–62; encourages
    children, 266–7; and Gore, 269
Fawcett, Henry, 5, 8
Fedden, Robin: on Crawford, 125;
    background, 166–7; as historic
    buildings secretary, 166–8, 271; on
    Ham and Osterley, 177; on NT in
    Wales, 182; in Northern Ireland, 187;
    threatens resignation over Smith, 203;
    and Fawcett, 259; co-edits NT Guide,

263; opposes admission of children, 266–7; Gore succeeds, 269; and Erddig, 275; *The Continuing Purpose*, 167

Fell and Rock Climbing Club, 43

Fenton House, Hampstead, 273

Ferguson's Gang, 70–1

Ferniehirst Castle, Roxburghshire, 78

Field, Ellen, 209

Field, The (magazine), 169

Fife fishing ports, 67

Finance & General Purposes Committee, 50

Finance Committee (NT), 244–6, 311–12

Flatford Mill, Suffolk, 118

Florence Court, County Fermanagh, 184, 187

Forest Park, Argyllshire, 68

Forestry Act (1947), 195

Forestry Commission, 67–9, 195, 204, 291

Forsyth, W. A., 83

Foundation for Art (NT), 312

Fountains Abbey, Yorkshire, 268, 295

Fowler, John, 271–5

Fowler, Peter, 311

France: and country houses, 81–2

Friar's Crag, Cumbria, 45

Frizzell Homeguardian Insurance Scheme, 310

Fulking escarpment, Sussex, 320

Garden Fund (NT), 158

Garden History Society, 211

Gardens Committee (NT), 157–9, 208–10

gardens and gardeners, 157–60, 206–12; advisers, 209, 213

Gascoigne, Major-General Sir Julian, 247–8

Gathorne-Hardy, Robert, 135

Gaze, John, 126, 132, 251

General Enclosure Act (1845), 19

General Purposes Committee (NT), 199, 201–2, 226, 246

Gentleman, David, 311

Georgian Group, 51, 83, 171–2

Giant's Causeway, County Antrim, 187, 221

Gibbons, Grinling, 146

Gibbs, Christopher, 89, 113, 119–20, 124, 197, 211, 219

Gibson, Patrick, Baron: and gardens advisers, 213; on Neptune Committee,

228; on Benson Committee, 240; recommends Bishop, 250–1; succeeds Antrim as chairman, 256–7; qualities, 257; interviews Fawcett, 259; and historic buildings representatives, 269; supports Stirling's appointment, 297; and Bradenham dispute, 299–301

Gill, Eric, 115

Gladstone, William Ewart, 1

Glanusk Park, Wales, 180

Glencoe, Argyllshire, 65–6

Glenfinnan Monument, Inverness-shire, 65

Golden Cap, Dorset, 224, 329

Goldfinger, Erno, 323

Goodhart-Rendel, Hal, 102

Gore, St John ('Bobby'), 269–71, 274

Gowbarrow Fell, Ullswater (Cumbria), 33

Gower, Lord Ronald, 150

Gower Peninsula, Pembrokeshire, 224

Gowers, Sir Ernest: Committee on preservation of country houses, 151, 156, 165, 184, 279

Grafton, Hugh Denis Charles FitzRoy, 11th Duke of (*earlier* Earl of Euston), 166, 173, 247

Grange Fell, Cumbria, 33

Grant, Duncan, 99

Gray, Mike, 324

Gray, Sylvia, 248, 264

Great Chalfield Manor, Wiltshire, 98

Great Gable, Cumbria, 43

Great Langdale, Cumbria, 205, 315

Great War (1914–18), 40–3

Great Wood, Derwentwater, Cumbria, 45–6

Green Belt (London and Home Counties) Act (1938), 104

Greenlands estate, Henley-on-Thames (Buckinghamshire), 109

Greg, Alec, 111

Greg, Samuel, 111

Greville, Mrs Ronald, 101, 146

Grey of Fallodon, Edward, Viscount, 55–6

Grigg, John, 230–2, 234

Grimond, Jo (*later* Baron), 132

Guardian (newspaper), 230–1, 234

Gubbay, Hannah, 271

Guiting Power (village), Gloucestershire, 232

Gunby Hall and estate, Lincolnshire, 100–1, 322

Hackney, London, 323–5
Haddington, George Baillie-Hamilton,
    12th Earl of, 65
Hadrian's Wall, Northumberland, 295
Hafod, Dyfed, 180
Hallett, Ray, 262
Ham House, Surrey, 144, 148–9, 155, 177,
    210
Hambleden, William Henry Smith, 3rd
    Viscount, 109
Hamer, Samuel H., 40, 47–8, 55, 57,
    59–60, 107
Hamer, Sir William, 33
Hampstead Heath, London, 20
Hanson-Smith, Christopher, 261
Harcourt, Sir William, 18
Hardcastle Crags, Yorkshire, 195
Hardwick Hall, Derbyshire, 169–70, 188
Hardy, Kate, 45n
Hardy, Thomas, 45
Harewood, Henry George Charles
    Lascelles, 5th Earl of, 123
Harewood House, Yorkshire, 156
Harkness, Edward, 72
Harlech, William Richard Ormsby-Gore,
    2nd Baron, 26
Harpur-Crewe estate, 281–3
Harpur-Crewe, Charles and Henry, 282
Hartley, L. P.: The Go-Between, 44
Harvey, John, 289, 293
Hatchlands, Surrey, 102
Hatfield Forest, Essex, 290–1
Hayter, Sir William, 238, 240
Head, Anthony, 1st Viscount, 247
Hearn, Katherine, 291
Hearth (Northern Ireland buildings
    preservation organization), 323
Heath, (Sir) Edward, 251, 253
Heelis, Mrs William see Potter, Beatrix
Heelis, William, 59
Helme, Tom, 274
Hemsley, J. H., 288
Hendref (Welsh buildings preservation
    trust), 323
Herbert, Robin, 247
Herbert, Sidney Charles, (later 16th Earl
    of Pembroke), 117
Herdwick sheep, 59–60, 205
'Heritage Coasts', 239
Heveningham Hall, Suffolk, 329
Hidcote, Gloucestershire, 158, 207, 210,
    212

Hill Farming Act (1945), 161
Hill, Caroline (Octavia's mother), 5
Hill, James (Octavia's father), 5
Hill, Captain John Burrow, 189
Hill, Miranda (Octavia's sister), 8, 21
Hill, Octavia: as co-founder of NT, 1–2,
    13, 18, 20–2, 25, 42; life and career,
    5–7; housing work and social reform,
    6–8, 15; defends Lake District against
    desecration, 10–11; and Rawnsley,
    11–12; qualities, 13–14, 39; and CPS,
    18; at Provisional Council meeting, 24;
    and restoration of Alfriston clergy
    house, 27, 304; on opening of
    Brandelhow Park, 31–2; and Princess
    Louise, 32; and countryside
    acquisitions near London, 33–4; on
    early acquisitions, 35; death, 39; and
    Harriot Yorke, 95; advises against
    acquiring Pollington, 110; preserves
    Toy's Hill, 151; and NT reforms, 246
Hill Top, Sawrey (Cumbria), 59
Hills, Ivan, 288
Hindhead Common, Surrey, 34, 265
Hinksey, Oxford, 16
Historic Buildings Committee (NT), 128,
    170, 178, 210; functions, 201, 244
Historic Buildings Councils: proposed,
    154, 156; established, 165–6; assistance
    for NT, 167–9, 208, 278; activities, 281,
    283
Historic Houses Association, 81, 278
Historical Buildings and Ancient
    Monuments Act (1953), 165
Hoare, Sir Henry and Lady, 90–1
Hobhouse, Sir Arthur, 140, 143, 160
Hobhouse, Hermione, 313
Hodgson, John, 267–8
holiday cottages, 264, 310
Holland-Martin, Admiral Sir Deric, 247
Holland-Martin, Edward ('Ruby'), 128,
    225, 228, 254
Holmes, F. A., 107
Holnicote estate, Somerset, 117–18
Hopper, Thomas, 178
Horan, Damaris, 306
Hornby, James, 24
Hornby, Sir Simon, 304, 314
Horne & Birkett, Messrs (solicitors), 129
Horne, Benjamin, 130
Horner Wood, Somerset, 118
Horsey, Norfolk, 287

Housesteads, Northumberland, 57–8
Hovingham, Yorkshire, 311
Howard, George, Baron, 135
Howarth, David, 274
Hudson, Edward, 101
Hughenden Manor, Buckinghamshire, 150, 266, 270
Hunt, Tiffany, 309
Hunter, Dorothy (Sir Robert's daughter), 4, 95
Hunter, Ellen, Lady (*née* Cann), 4
Hunter, Sir Robert: as co-founder of NT, 1–2, 13–14, 18, 20–1, 42; life and career, 2–5; meets Octavia Hill, 7; defends Lake District, 11; qualities, 13–14, 41; and CPS, 18; drafts NT constitution, 22, 24–5, 65; at Provisional Council meeting, 24; legislative reforms, 29–31; and acquisition of Hindhead Common, 34; death and tributes, 39; tribute to Octavia Hill, 39; regard for Bailey, 46; and founding of Ramblers' Association, 63; chairmanship of NT, 307; *The National Trust: Its Aims and Its Work*, 28
hunting: Beatrix Potter forbids by otter hounds or harriers, 60; abolition question, 314, 316–18
Hussey, Christopher, 75, 83, 166
Hussey, Trevor, 299
Hutchinson, Professor George, 299–300, 303
Huxley, Thomas Henry, 18, 24

Ickworth House, Suffolk, 206
Ilam Hall, Dovedale (Derbyshire), 107
India: National Trust for Art and Cultural Heritage, 330
industrial archaeology, 190–3
Inglewood, William Morgan Fletcher Vane, 1st Baron, 69, 166
insurance, 310
International Castles Institute, 306
International Conference of National Trusts, First (Culzean Castle, Ayrshire), 67; Second (USA), 67
Iona (island), 67
Ireland, Phyllis, 293–4
Isaacson, Wootton, 160
Iveagh, Gwendolen Florence Mary, Countess of, 271

Jackson-Stops, Gervase, 274–6, 281, 306
Jaffé, Michael, 270
Jekyll, Gertrude, 36
Jersey, George Francis Child-Villiers, 9th Earl of, 149
Jessel, Edward Herbert Jessel, 2nd Baron, 216
Joekes, Rosemary, 263
John Lewis Partnership, 190
Johnson, Samuel, 107
Johnston, Lawrence, 158
Jones, Arthur Creech, 123
Jones, Colin, 285
Jones, Tom, 109
Joseph, Sir Keith, 223
Jowett, Benjamin, 9
Jubilee Appeal (NT), 136, 163

Kedleston Hall, Derbyshire, 53–4, 283, 308
Keeling, Edward, 128, 141
Keswick, Cumbria, 11; School of Industrial Arts, 13
Keynes, John Maynard, Baron, 82
Killarney, Lakes of (Ireland), 28
Killerton estate, Devon, 117–18, 161
Killynether House, County Down, 112
Kinder Scout, Derbyshire: 1932 mass trespass, 106–7; acquired by NT, 108; nature conservation, 292
King's How, Cumbria, 33
Kingston Lacy, Dorset, 296
Kinver Edge, Staffordshire, 195
Knole House, Kent, 75, 87–8, 90, 145, 154, 167, 253
Knollys, Eardley, 99, 132, 163, 194, 220
Kynance Cove, Cornwall, 108
Kyrle Society, 21, 24, 32

Labour Party: postwar government and legislation, 122, 135–8, 155–6, 161; and listed buildings, 165
Lacock Abbey, Wiltshire, 94, 96, 163, 167
Lacock (village), Wiltshire, 111
Ladies' Guild, 6
Lake District: Rawnsley's activities in, 9–11, 22; as National Park, 10, 141, 318; NT acquisitions in, 31–3, 43, 45–6, 112, 118, 120, 143, 160, 189, 204; regional administration for, 50, 112, 198; Trevelyan's interest in, 58; Beatrix Potter and, 59; and conifer

plantations, 68; land cultivation, 116; farm improvements and management, 161, 204–5; camping, 205; cars, parking and caravan sites, 205; nature conservation in, 288; appeal for, 314–15; landscape and building survey, 315, 321; footpath damage and 'pitching' repairs, 318–19; *see also* Acland, Cuthbert H. D.

Lake District Defence Society, 10

land agents, 285

Land Fund *see* National Land Fund

land, inalienability of, 31, 139–40, 215–16, 302, 328

land sales: inter-war, 43–4

land use and development: legislation on, 138–9

landed estates (and country houses): under threat, 43–4, 75–6, 79–80; listed for merit, 83; postwar cost of upkeep, 143; Gowers' Committee on, 151–4, 156; opened to visitors, 173–4; postwar acquisition diminishes, 206; artistic achievements, 332

Landmark Trust, 191

Lang, Brian, 309

Langdale Pikes, Cumbria, 204

Lankester, Patricia, 268

Lascelles, Sir Alan, 166

Latham, Robert, 232

Lauderdale, Elizabeth Murray, Duchess of, 148

Laurence, Philip, 3

Lawrence, Sir Alexander, 83

Lawson, Nigel, Baron, 283

League Against Cruel Sports, 316

Learmont, Mrs J. B., 41

Leconfield, Charles Henry Wyndham, 3rd Baron, 33, 103, 146

Leconfield Commons, Cumbria, 204, 295

Lees-Milne, Alvilde, 208–9

Lees-Milne, James: on Zetland, 77; on Matheson, 81; as secretary of Country Houses Committee, 83–6, 97–9, 102; background and career, 84; appointed historic buildings adviser, 85; admires Esher, 89; seeks to persuade owners to transfer properties to NT, 89, 103; on Hoares at Stourhead, 91; on Matilda Talbot, 94, 96; wartime work for NT, 97–8; travels and visits for

NT, 99; and transfer of Gunby Hall, 101; on acquisition of Swanton Morley, 119; duties defined 124; on Holland-Martin, 128; under Crawford's chairmanship, 129; relations with Mallaby, 130; on Bevir, 131; on Harold Nicolson, 131; welcomes Rathbone's appointment, 132; on area representatives, 134; on Esher's view of bureaucrats, 142; on wartime vandalism in country houses, 144; on quality of acquisitions, 145; and Vita Sackville-West's guide to Knole, 145; visits Lyme Park, 147; on Ham House, 148–9; Lord Jersey praises, 149; drafts memorandum for Gowers Committee, 152; and government grants, 155; recommends Hidcote, 159; and postwar openings, 162; on Stourhead management, 163; Fedden succeeds as historic buildings secretary, 166–7; report on Penrhyn castle, 178; and Lady Mairi Bury, 186; and Cook estates, 189; and Trinick's work in Cornwall, 220; on unwieldiness of Properties Committee, 244

Leighton, Sir Frederic (*later* 1st Baron), 24

Leoni, Giacomo, 147

Levy, Mr & Mrs Kenneth, 266

Lincoln, Abraham, 119

Lindisfarne Castle, Northumberland, 101

Lindsay, A. D., 1st Baron Lindsay of Birker, 123

Linnean Society, 24

Little Houses Improvement Scheme (Scotland), 67

Liverpool City Council, 100

living standards: 1960s improvements, 199

Lizard Peninsula, Cornwall, 108

Llangibby Castle, Monmouthshire, 180

Llanwern Park, Monmouthshire, 180

Lloyd, George Ambrose Lloyd, 1st Baron, 85

Lloyd George, David, 1st Earl, 30, 61, 78, 136

Lloyd-Johnes, Revd Herbert, 114

Llyn Peninsula, Gwynedd, 320

Local Committee Conference (NT) (1949), 68–9

Lodore Falls, Cumbria, 22

Londonderry, Charles Stewart Henry
    Vane-Tempest-Stewart, 7th Marquess
    of, 185
Londonderry, Edith, Marchioness of, 186
Long Crendon Court House,
    Oxfordshire, 35
Longleat, Wiltshire, 85, 174, 196
Lonsdale Commons, Cumbria, 204
Lords Island, Cumbria, 45
Lothian, Philip Kerr, 11th Marquess of:
    speech on country house
    preservation, 78–82, 86, 90, 160; in
    1939 National Trust Bill debate, 88;
    influence on Charles Trevelyan, 93;
    death, 98
Louise, Princess, Duchess of Argyll: as
    president of NT, 31–2, 123
Lubbock, Cecil, 128
Lutyens, Sir Edwin, 32, 101
Lyle, Colonel A. A., 35
Lyme Park, Cheshire, 147, 163, 206
Lyveden New Bield, Northamptonshire,
    51

McCausland, Marcus, 112–13
McCracken, Leslie, 329
McDermott, Bernie, 315
MacDonald of Glenaladale, Alexander, 65
MacDonald, James Ramsay, 52, 56, 61–2,
    69–70
MacDonald, John, 300
MacDougall, Sir Robert, 107
McIntyre, James, 187
Macmillan, Harold, 1st Earl of Stockton,
    138, 199
Mallaby, (Sir) George, 130–1, 134, 142
Malvern Hills, 20
Man, Calf of, 113
Man, Isle of, 113
Manchester Corporation, 217
Mander, Sir Geoffrey, 92
Mander, Rosalie, Lady, 92
Manifold Trust, 191
Manifold Valley, Derbyshire, 105, 107–8
Manx Museum and National Trust, 113
Mariner's Hill, Kent, 151
Marlborough, Charles Richard John
    Spencer-Churchill, 9th Duke of, 75–6
Marlborough, John Albert Edward
    William Spencer-Churchill, 10th Duke
    of, 196
Marsden Moor Estate, Yorkshire, 189

Martineau, Anthony, 129–30, 159, 170
Mary, Queen, 123–4
Massachusetts: Trustees of Public
    Reservations, 22, 24
Matheson, Donald Macleod: relations
    with Zetland, 77; qualities, 80–1; and
    Country House Scheme, 83; and
    Trevelyan's gift of Wallington, 92; and
    Lacock Abbey, 94; in Second World
    War, 97; illness, 98; and Speke Hall,
    100; paper on Country-House
    Scheme, 101; and Northern Ireland,
    112, 182; and Lloyd-Johnes, 114; on
    NT management and financial
    weaknesses, 120, 133; appoints
    Martineau, 129; and National Parks,
    142; and relations with Cook, 189
Maugham, Frederick Herbert Maugham,
    1st Viscount, 88
Maurice, F. D., 6
Maxwell, Sir John Stirling, 65
Meadfields, Haslemere, Surrey, 4
Mears, Frank, 64
Medway Queen (ship), 224
Melford Hall, Suffolk, 83
Mentmore Towers, Buckinghamshire,
    278–9
Merseyside Metropolitan County, 100
Merthyr, William Brereton Couchman
    Lewis, 3rd Baron, 178, 181
Merton Abbey Wall, 49
Merton Borough Council, 325
Merton, Statute of (1235), 19
Messel, Colonel Leonard, 171
Messel, Oliver, 171
Methuen, Paul Ayshford Methuen, 4th
    Baron, 83, 88
Metropolitan Commons Act (1866), 20
Midland Bank affinity card, 310
Mill, John Stuart, 18
Minchinhampton Commons,
    Gloucestershire, 293
Mitchell, Lieutenant-Colonel, 289
Mitchell, Peter, 289, 323
Mlinaric, David, 274
Monk Coniston estate, Cumbria, 60
Monmouth, 114
Montacute, Somerset, 53, 72, 86–7, 152,
    162, 189; garden, 206, 211
Montagu of Beaulieu, Edward John
    Barrington Douglas-Scott-Montagu,
    3rd Baron, 174, 255

Montgomery-Massingberd, Diana, Lady, 101
Monuments and Buildings (Protection) Bill (1951), 155
Moorman, Mary (née Trevelyan), 231
Morden Hall estate, London, 118, 325
Morris, (Sir) Parker, 95n
Morris, William, 15, 17, 21, 26, 92, 171
Morte Point, North Devon, 38
Moss, Martin, 263
motor cars: effects of, 104–5, 199, 205, 220; see also road building
Mount Stewart, County Down, 183–5, 211
Mowat, Robert, 231, 236, 248
Museum of Childhood, Sudbury Hall, Derbyshire, 267
Mussenden Temple, County Londonderry, 187

Nash, John, 118
National Art Collections Fund, 27
National Association of Social Science, Birmingham meeting (1884), 20
National Coal Board, 275–6
National Footpaths Preservation Society, 29
National Gardens Scheme, 158–9
National Heritage Memorial Fund: replaces Land Fund, 138, 279; makes endowments and grants, 165, 280–4, 295–6; buys Nostell Priory collection for NT, 170, 283; functions and operations, 279–80
National Land Fund: established, 30, 136–8; replaced by National Heritage Memorial Fund, 138, 279; effect on offers of country houses, 143, 156; aids NT, 189–90, 204, 208, 221, 278, 295, 297
National Parks: proposed, 62–3, 104, 135, 139–40, 286; established, 141–3, 156
National Parks and Access to the Countryside Act (1949), 139, 141, 286
National Parks Commission, 140
National Plant Collections, 160
National Trust: formally established, 1, 21–3, 25; name, 21; Provisional Council, 24; acquires first property (Dinas Oleu), 25–6; articles of association, 25; early financial struggle, 27–8; initial aims and policy,
28; parliamentary legislation on, 30–1, 36, 80–1, 87–8, 115, 215, 248–9, 287; and inalienability of land, 31, 139–40, 215–16, 302, 328, 333; cash donations, 40, 105; membership numbers and recruitment, 40, 45, 52, 69, 115–16, 197, 242, 259–61, 265, 308–9, 334, 337; properties and sites, x, 40, 253; subscription revenues, 40, 69, 115–16, 265, 309; seeks to define 'historic interest' and 'natural beauty', 42; financial management and control, 49–50, 195, 242, 245; selection standards for acquisitions, 49, 71; regional representatives, 50, 133; volunteer helpers, 50, 263–5; archaeological and historic acquisitions, 51–2; avoids governmental commitments and subsidies, 63, 152–3; links with countryside associations, 63; coastal properties and preservation, 66–7, 108–9, 199, 219–21, 238–9, 319–20; legacies and donations, 69, 195, 266, 303; social background of officers, 95–6, 247–8, 303; land gifts, 105; as landlord, 110; village acquisition, 110–11; regional committees, 112, 246–8; general endowment and reserve funds, 115–16, 163; land acreage (owned and covenanted), 115–16, 160, 204, 253, 295, 308, 337; membership fee, 115; oakleaf symbol adopted, 115; total income, 116, 163, 253, 308; postwar role and operations, 120–3; postwar changes and appointments of officers, 123–31; legal advisers, 129–30; secretaryship, 130–1; area agents, 133–5; historic buildings representatives, 134–5, 269; concern for artistic matters, 134–5; Jubilee Appeal, 136, 163; and Land Fund, 136–8, 143, 189–90, 204, 221, 278, 297; and National Parks, 141–3, 156; postwar property acquisitions, 145–51; and Gowers Committee, 152; accepts maintenance grants from government, 69; and gardens, 157–60, 206–13; staff numbers, 161, 194, 197, 258, 337; condition of properties, 162–3, 196; numbers of visitors to properties, 162, 195, 212, 253, 334;

financial problems and deficits, 163–4, 194, 253–4; and industrial archaeology, 190–4; investment policy, 194; income from admissions to properties, 195; local authorities help, 195; timber sales, 195; press and public criticisms of, 196–7; administrative and management reforms, 197–8, 200–2, 249, 251–2, 304; regional secretaries and directors appointed, 198; administration criticized by Conrad Rawnsley, 224–35, 239; access complications, 232–3; 1967 Committee of Enquiry into reforms, 237–8; Benson report and recommendations on, 238, 240–52, 259, 304; constitution, 242, 249; voting procedures, 242; committee structure and functions, 244–6, 252, 304–5, 310–11; panels appointed, 244; accounting practice and presentation, 248–9; staff salaries, wages and pensions, 251, 253; staff relations with committees, 252, 304; total expenditure, 253; young people engaged under Government Employment Schemes, 256; public relations and events, 258–62, 309; subscriptions collection arrangements, 260–1, 266; regional information officers, 261; shops and trading activities, 262–4; publications, 263–4; catering 264; holiday cottages, 264, 310; volunteers, 265; fund-raising, 266; policy and practice on children and education, 266–8; and nature conservation, 36–7, 285–93; Arkell committee of investigation into, 302–3, 310; relationship with members, 302, 310; conservatism, 303; Hornby Committee on, 304–5; computer system renewed, 309–10; women in senior positions, 309; conduct of business, 311–15, 331; vernacular buildings survey, 321; relations with government and other bodies, 327–8; opposes farm land deregulation and water privatization, 328; international links, 329–30; role and future, 331–4; *see also* Council; Extraordinary General Meetings; and individual committees

National Trust Acts: (1907), 30–1, 36, 115, 215, 287; (1937), 80–1, 87–8; (1939), 88; (1953), 115; (1971 Bill), 249
National Trust Bulletin, 188
National Trust Centres, 265
National Trust Guide (ed. Robin Fedden and Rosemary Joekes), 263–4
National Trust for Scotland: formation and activities, 64–7, 321; aid from Pilgrim Trust, 72; and nature conservation, 288; and oil industry, 333
Natural Environment Council, 214
Natural History Committee *see* Advisory Committee on Natural History
Nature Conservancy, 190, 214, 289, 292
Nature Conservancy Council, 286, 288
Nature Conservation Review, A, 286, 288, 291
nature reserves, 36–7, 286, 288–9, 293, 296
Neptune, Enterprise (appeal), 67, 199, 217, 219, 222–3, 227–8, 230, 233–4, 238–9, 265, 319; Silver Jubilee Appeal (1990), 320
New Towns Act (1946), 138
Newbattle Castle, Midlothian, 78
Newgale, Pembrokeshire, 109
Newman, Jan, 38
Newton, Richard William Davenport Legh, 3rd Baron, 147
Newtown, Isle of Wight: old Town Hall, 71
Nicholson, Max, 219
Nicolson, Sir Harold, 123, 128, 131, 143, 152–3, 165, 200, 207, 249
Nicolson, Nigel, 90, 207–8
Noailles, Duc de, 81
Noel-Buxton, Noel Edward, 1st Baron, 71
Norman, Mark: on Rathbone, 133; on Antrim, 218; serves on Neptune Committee, 228; as chairman of Finance Committee, 244–5, 254; on Anthony Head, 247; on Fawcett, 259; recommends Stirling, 297
Norman, Ronald C.: chairs Finance & General Purposes Committee 49; declines NT chairmanship, 76, 124; serves on Executive Committee, 83; Esher succeeds, 96–7; devotion to NT, 129, 244; on acquisition of country houses, 152
Northern Ireland: and little houses

scheme, 67; regional committee, 112–13, 182–3, 186, 198, 247; and Land Fund, 138, 183–6, 280; properties and acquisitions, 183–6; coastline protection, 221; *Guide to Properties*, 187
Norwich, John Julius Cooper, 2nd Viscount, 311
Nostell Priority, Yorkshire, 170, 210, 283
Nunnington, Yorkshire, 210
Nymans, Sussex, 171

Observer (newspaper), 229
Ogwen, Lake, Caernarvonshire, 178
Oliver of Aylmerston, Peter Raymond Oliver, Baron, 241, 249, 318
Onslow, William Arthur Bampfylde Onslow, 6th Earl, 271
Open Spaces Society, 297
Orde, Major Peter, 247
Osterley Park, Middlesex, 148–50, 155, 177

Packwood, near Birmingham, 97, 206
Parliamentary Committee on Forests, Commons and Open Spaces in and around the Metropolis (1865), 19–20
Paton, Major Hadden, 244
Paxton, Sir Joseph, 278
Peak District, Derbyshire: NT interest and possessions in, 105, 107–8, 143, 266; 1932 'trespass', 106–7; as National Park, 106, 108, 141; land cultivation, 116; volunteer helpers in, 265
Peek, Sir Henry, 3
Peers, Sir Charles, 83
Pembroke Coastal Appeal, 219
Pennine Way, 107
Penrhyn Castle, Gwynedd, 137, 178, 180–1, 188
Peto, Major Michael, 83
Petworth House, Sussex: acquired by NT, 33, 102–3, 146; contents transferred to NT, 138, 146, 154; reduction in servants, 144; Duke of Bedford criticizes, 173; proposed bypass through park, 216–17
Pevsner, Sir Nikolaus, 259
Philip, Prince, Duke of Edinburgh, 203, 222–3, 320
Picton Castle, Pembrokeshire, 180
Pilgrim Trust, 67, 72, 107, 109, 118, 223
Pitt-Rivers, General Augustus, 51

Plas Newydd, Anglesey, 166
Plymouth, Robert George Windsor-Clive, 1st Earl of, 40–1, 46–7
Polesden Lacey, Surrey, 101, 146, 167
Pollington, Northamptonshire, 110
Pope-Hennessy, Sir John, 270
Popham, John 299
Potter, Beatrix (Mrs William Heelis): as NT benefactor, 59–60; as activist, 107, 261
Powis castle, Powys, 180–1, 280
Powis, George Charles Herbert, 4th Earl of, 180–1
Preece, Roy, 263
Prideaux, Julian, 299, 308, 325
Priestley, J. B., 116–17
Princess Royal *see* Victoria Alexandra Alice, Countess of Harewood
Prior Park, Bath, 216
Properties Committee (NT), 85, 244, 246, 311
Publicity Committee (NT), 114, 128
Pullen, David, 313
Punch (magazine), 25
Putney Heath, 19

Quantock Hills, Somerset: red deer, 317
Quebec House, Westerham (Kent), 41
Queen Anne's Gate, London: NT moves head office to, 123
Queen's Institute for District Nurses, 158

Rackham, Oliver, 290–1
Radcliffe, Cyril John, Viscount, 237
Rainbow Wood Farm, Bath, 215–16
Rainham Hall, Essex, 137
ramblers: access to countryside, 105–7
Ramblers' Association, 63, 140, 297
Rathbone, J. W. F.: appointed secretary, 132; qualities, 132–3; and National Parks, 141; and Powis Castle, 180; and Northern Ireland, 185–7; and Brownsea Island, 190; and NT finances, 194; criticizes administration, 197–8; Trinick writes to on administrative reforms, 201; under criticism, 202–3; and appointment of director of Neptune, 222; conflict with Conrad Rawnsley, 224–7, 235, 239; doubts on Enterprise Neptune, 225; and 1967 postal vote, 237; nervous collapse and retirement,

239, 249–50, 255; sends minutes to Clark, 243; and Hatfield Forest, 291

Ravensbourne Registration Service, Beckenham, 260

Rawnsley, Commander Conrad: disputes with and criticisms of NT, 217, 224–7, 232–4, 239, 248; as director of Enterprise Neptune, 222–4, 228; character and personality, 224–5; given notice, 228–9, 239; public attacks on NT, 229–30; calls Extraordinary General Meeting, 231, 233, 235–7, 317; speaks at 1966 AGM, 231; hip operation, 238; makes homosexuality accusations to Benson Committee, 242

Rawnsley, Edith (*née* Fletcher; Hardwicke's first wife), 9–10, 12–13

Rawnsley, Eleanor (*née* Simpson; Hardwick's second wife), 13

Rawnsley, Canon Hardwicke: as co-founder of NT, 1–2, 13, 22; life and career, 8–10; and Lake District, 9–11, 22, 31, 33, 45–6; travels and writings, 10, 13; declines bishopric of Madagascar, 11–12; supports variety of causes, 12, 107; visits USA, 12; canonry of Carlisle, 13; death, 13, 33, 45; qualities, 14; appointed honorary secretary, 25; on Dinas Oleu, 25–6; sonnet on Hunter, 39–40; Beatrix Potter meets, 59; and Scotland, 64

Reconstruction, Ministry of, 42

Rees, Goronwy, 181

Reiss, Mrs F. E., 211

Repton, Humphrey, 102, 312

Restriction of Ribbon Development Act (1935), 104

Rice, Ladislas, 304

Ridley, Nicholas, Baron, 232, 246, 329

road building and development, 216, 329

road traffic, 104–5

Robbins, Lionel Charles, Baron, 238

Roberts, Anne, 309–10

Roberts, Goronwy, 181

Robinson, (Sir) Kenneth, 181, 225

Robinson, Sir Roy, 68

Rogers, Michael, 277

Roper, Lanning, 209

Rosebery, Neil Archibald Primrose, 5th Earl of, 24

Rosse, Anne, Countess of (*née* Messel), 171–2

Rosse, Laurence Michael Harvey Parsons, 6th Earl of: on Crawford's aesthetic sense, 126; serves on Executive Committee, 128; proposes area representatives, 134; and gardens, 157–8; and Hardwick Hall, 169; as chairman of Historic Buildings Committee, 170–1; background, 170, 183; visits Northern Ireland, 187; lacks interest in administration, 198; and appointment of John Smith, 200; and Rathbone, 202; and George Taylor, 209; doubts on Antrim as successor to Crawford, 217; and Enterprise Neptune, 225; and committee reforms, 244; and Rathbone's retirement, 250; and appointment of Boles, 256; and Gibson's appointment as chairman, 256–7; interviews Fawcett, 259

Rothman, Benny, 106–7

Rothschild, Anthony de, 38, 135

Rothschild, Baron Ferdinand de, 174

Rothschild, Mr and Mrs James de, 38, 175–6

Rothschild, Baron Meyer de, 278

Rothschild, Miriam, 36, 287

Rothschild, Nathaniel Charles, 36–8

Rough Fort, Northern Ireland, 113

Rousham, Oxfordshire, 84

Rowallane, County Down, 184

Royal Academy of Arts, 24

Royal Botanic Society, 24

Royal Horticultural Society: garden rescue scheme, 157

Royal Oak Foundation, 305

Royal Society of Arts, 110

Royal Society for Nature Conservation (*formerly* Society for the Promotion of Nature Reserves), 37

Royal Society for the Protection of Birds, x

Runnymede, 73; NT wartime headquarters in, 96

rural life: changes in, 44–5

Ruskin, John: and Octavia Hill, 6, 8; and Rawnsley, 8–9, 12–13; activities, interests and qualities, 15–17, 23; and SPAB, 15; and Fanny Talbot, 25; enthusiasm for medieval architecture, 26; at Wallington, 93; *Seven Lamps of Architecture*, 16

Russell, Arthur, 65
Russell, George S.: *The Formative Years, 1929–1939*, 64n

Sach, Margaret, 119
Sackville, Major-General Charles John Sackville-West, 4th Baron, 90, 123, 145–6
Sackville-West, Edward (*later* 5th Baron Sackville), 145
Sackville-West, Vita: and Lees-Milne, 85; social circle, 123; and transfer of Knole to NT, 145; on Gardens Committee, 158; creates Sissinghurst garden, 207–8; opposes appointment of George Taylor, 209; designs Montacute garden, 211
St David's Head, Pembrokeshire, 109
St Kilda (island), 67
St Oswald, Rowland George Winn, 3rd Baron, 170
Sales, John, 213
Salisbury, Sir Edward, 158, 287
Salisbury, James Edward Hubert Gascoyne-Cecil, 4th Marquess of, 83
Salisbury, Robert Arthur Talbot Gascoyne-Cecil, 3rd Marquess of, 1
Saltram Park, Plymouth, 216, 220, 229
Sambourne, Linley, 172
Sandwith, Hermione, 270
Save Britain's Heritage, 282
Scafell Pike, Cumbria, 33
Scarsdale, Francis John Nathaniel Curzon, 3rd Viscount, 283
Scarsdale, Richard Nathaniel Curzon, 2nd Viscount, 54
schools: visits from, 268
Scolt Head, Norfolk, 289
Scotland, 63–7; *see also* National Trust for Scotland
Scotney Castle, Kent, 75
Scott, Sir George Gilbert, 172
Scott, Sir Giles Gilbert, 17
Scott, Peter, 247
Scott Report on Land Utilization in Rural Areas, 138
Scott-Elliot, Walter, 129
Scottish Heritage USA, 67
seal culling, 199, 213–14, 217
seashore properties *see* coastal (seashore) properties and acquisition
Selworthy Beacon, Somerset, 118

settled land, 88
Seven Sisters (cliffs), Sussex, 73, 220
Shalford Mill, Surrey, 70
Shaw, Norman, 172, 277
Shaw-Lefevre, George John (*later* Baron Eversley), 3, 18–22, 24
Sheffield Park, Sussex, 208, 211, 326
Sheldon, Paul, 316
Sheringham Park, Norfolk, 312
Sherwood Forest, Nottinghamshire, 118
Shuckburgh, Sir Evelyn, 248
Shugborough, Staffordshire, 206, 271
Signal Rock, Glencoe, 65
Silkin, Lewis, 1st Baron, 127, 138–41
Sissinghurst, Kent, 90, 207–8
Sites of Special Scientific Interest, 286, 288
Six Essays on Commons Preservation, 3
Skiddaw (mountain), Cumbria, 11
Slindon estate, Sussex, 160, 293, 326
Smith, Hubert, 119–20, 130, 160, 197
Smith, (Sir) John: and Stratford-upon-Avon canal, 190–3; and changing lifestyles, 199; reforms as chairman of General Purposes Committee, 199–202; relations with Rathbone, 202–3; and Conrad Rawnsley, 222, 239; serves on NT Council, 314
Smith, Dr Thomas Southwood, 5
Snowdon (mountain), Wales, 29
Snowdonia (Wales), 114, 178, 188; designated a National Park, 141
Soane, Sir John, 277
Society for the Promotion of Nature Reserves *see* Royal Society for Nature Conservation
Society for the Protection of Ancient Buildings (SPAB): founded, 15–17; supports NT, 18, 24; and Alfriston clergy house, 26; and Stonehenge, 29; and Cook's gifts to NT, 72; 'anti-scrape' doctrine, 83; on Speke Hall, 100; and Georgian Group, 172
Souter Lighthouse, Tyne and Wear, 330
South Downs National Park (proposed), 160
South Shields, Tyne and Wear, 320
Spectator (journal): on Hunter, 39
Speke Hall, near Liverpool, 100
Spencer, Albert Edward John, 7th Earl, 97
Stackpole, Pembrokeshire, 297
Stainton, Sheila and Sandwith, Hermione: *The Manual of Housekeeping*, 270

Standen, Sussex, 277
Steel, Robert, 320
Steventon Priory, Oxfordshire, 71
Stirling, (Sir) Angus, 297–8, 303, 305, 308, 323, 324, 329
Stockport, Borough of, 147
Stoneacre, Kent, 71
Stonehenge, Wiltshire, 29, 52, 73, 162, 313
Stonehenge Down, Wiltshire, 293
storm disasters, 266, 326
Stourhead, Wiltshire, 90, 157, 161–3, 206, 211–12, 259, 262
Stowe, Buckinghamshire, 308
Strachey, St Loe, 110
Stratford-upon-Avon canal, 190–3, 265
Stratfield Saye, Berkshire, 102
Strauss, Harry, 141
Street, Arthur George, 104
Stubbins estate, Lancashire, 118
Studland Bay, Dorset, 297
Styal (village), Cheshire, 111
Stybarrow Crag, Ullswater, Cumbria, 33
Sudbury Hall, Derbyshire, 267, 271–2
Sugar Loaf, Abergavenny, 113
Summerson, Sir John, 259
Sunday Express, 196
Sutherland, Graham, 99
Sutton House, Hackney, 323–5
Swanton Morley (village), Norfolk, 119
Swiss Cottage Fields, London, 7
Symonds, H. M., 58

Talbot, Fanny, 25–6
Talbot, Matilda, 94–6, 111, 163
Talbot, William Henry Fox, 94
Tansley, Arthur, 289
Tarn Hows, Cumbria, 205
Tattershall Castle, Lincolnshire, 30, 52–3
Tatton Park, Cheshire, 188, 195
Tawney, Richard Henry, 123
taxation: effect on landed estates, 43–4, 75, 78–80; exemptions on NT properties, 69; bequests exempted from, 251; exemptions for outstanding historic buildings, 278; benefits from exemptions, 333
Taylor, George, 208–9
Taylor, Harold Victor, 158
Telford, Thomas, 194
Templer, Field Marshal Sir Gerald, 203
Tenby, Dyfed, 114

Tennyson, Alfred, 1st Baron, 8
Tetley, John, 178–9, 181
Thackray, David, 294
Thatcher, Margaret (later Baroness), 253
Thomas, Graham, 157, 209–13; Gardens of the National Trust, 263
Thomas, Sir Keith, 316
Thompson, Bruce, 60, 112, 133
Thorneycroft, Peter (later Baron), 138
Thring, Edward, 9
Tilberthwaite, Cumbria, 60
Times, The: on Hunter, 5; on 1st Duke of Westminster, 15; supports founding of NT, 23, 25; on Hardwicke Rawnsley, 45; NT appeals in, 52, 55–6; Zetland writes to on institutional occupancy of houses, 99; publicity for NT, 115; criticisms of Sir Richard Acland in, 117; Crawford writes letter to, 154; on opening arrangements for NT houses, 162, 232; and Conrad Rawnsley's criticisms of NT, 233–4
Tintagel, Cornwall, 35, 265
Tintinhull, Somerset, 211
Tollemache family, 148
Tollemache, Sir Lyonel, 148–9
Town and Country Amenities Act (1974), 210
Town and Country Planning Acts: (1931), 69, 104; (1947), 138–9, 172, 220
Toy's Hill, Kent, 151
'Treasure Houses of Britain' exhibition, Washington (1985–6), 305–6
Tredegar Park, Monmouthshire, 180
Treitel, Professor Guenter, 313
Trelissick, Cornwall, 212
Trengwainton, Cornwall, 207, 212
Tresham, Sir Thomas, 51
Trevelyan family, 81
Trevelyan, Sir Charles: transfers Wallington to NT, 92–3
Trevelyan, George Macaulay: on Bailey, 46; work for NT, 48, 56, 58, 61, 73, 96, 128; and Ashridge estate, 55–6, 313; on preservation of countryside, 56–7, 107; and Lake District, 58–9, 68–9; Beatrix Potter on, 60; as president of YHA, 63; opposes conifer plantations, 67–9; and Stanley Baldwin, 72; and transfer of Wallington to NT, 92; on Lees-Milne, 98; and restrictive covenants, 109; defends Sir Richard Acland, 117;

unable to attend meetings in term
time, 127; recommends Chorley for
NT committees, 128; retirement, 134;
Must England's Beauty Perish?, 57
Trevelyan, Mary, Lady, 92
Trinick, Michael: as area agent in
Cornwall, 189, 198, 220–1; urges
administration reforms, 201–2;
opposes Saltram park road scheme,
216; property acquisitions, 220; at 1967
Extraordinary General Meeting, 236;
opposes expenditure cuts, 254–5;
encourages promotion of NT, 261
Trollope, Anthony, 1
Troutbeck Park, Cumbria, 60

Ullswater, Cumbria: water extraction
proposals, 217
Ulster Architectural Heritage Society, 323
Ulster Coastline Appeal, 221
Ulster Land Fund, 138, 183–6, 280
United States of America: Rawnsley
visits, 12; national parks and nature
conservation, 22; corresponding NT
members in, 40; financial support for
NT, 305
Unna, Percy, 66
Upcott, Janet, 95, 246
Uppark, Sussex, 173, 275; fire damaged
and restored, 327
Upton House, Warwickshire, 135, 211
Urry, Audrey, 299–301, 303
Urry, Professor S. A., 299
Usk valley, Wales, 113
Uthwatt Report on Compensation and
Betterment, 138

Vallance, Aylmer, 71
Vane, William Morgan Fletcher *see*
Inglewood, 1st Baron
Vanishing Coast, The (film), 223
vernacular buildings, 321–5
Verney, Sir Ralph, 168–9
Vernon, John Lawrence Vernon, 10th
Baron, 272
Victoria and Albert Museum, 155, 177,
270; 'The Destruction of the English
Country House' exhibition (1975), 278
Victoria Alexandra Alice, Countess of
Harewood (Princess Royal), 123–4,
156
Victorian Society, 172

villages: acquisition of, 110
Viollet-le-Duc, Eugène Emmanuel, 17
Vyne, The, Hampshire, 188

Waddesdon Manor, Buckinghamshire,
38, 174–6
Wales: acquisitions and properties,
113–14, 177–80; separate committee
formed (Committee for Wales), 113,
181, 198; and Conrad Rawnsley's
tactlessness, 226–7; nature
conservation in, 288
Wallace, Carew, 198
Wallington, Northumberland, 56, 92–3, 97
wardens (countryside), 319
Warner, Sir Fred, 248
Warwick Castle, 174
Washington, DC *see* 'Treasure Houses of
Britain' exhibition
Washington, George, 195
Washington Old Hall, County Durham,
195, 305
Wastwater (and Wasdale), Cumbria, 204
Watermeads, south London, 325
Waterson, Merlin, 276
Watlington Hill, Oxfordshire, 118
Watlington Park, Oxfordshire, 97
Watson, David and Graham, 321
Watt, Adelaide, 100
Waugh, Evelyn: *Brideshead Revisited*, 174;
A Handful of Dust, 51
Webb, Philip, 277
Wellington, Gerald Wellesley, 7th Duke
of: on Country House Committee, 83;
and future of Apsley House, 102; on
Penrhyn castle, 178
Wemyss, Francis Charteris, 10th Earl of, 8
Wemyss, Francis David Charteris, 12th
Earl of, 210
West Penwith, Cornwall, 295
West Wycombe Park, Buckinghamshire:
NT acquires, 96, 98; NT occupies in
war, 96–7; condition, 163
West Wycombe (village),
Buckinghamshire: acquired by NT,
110–11
Westminster, Hugh Lupus Grosvenor, 1st
Duke of: as first NT president, 1–2, 8,
12, 14, 24, 32, 65; background and
activities, 14–15; supports CPS, 18;
prophesies successful future for NT, 25
Weston Park, Staffordshire, 283

Wharfedale, Yorkshire, 321
What to See (atlas), 263
Wheeler, Sir Mortimer, 293
Wheeler, R. J., 254
Which? (magazine), 199
Whistler, Rex, 166
Whistling Sands, Llyn Peninsula,
    Gwynedd, 320
White, Commander Harry, 184
White Park Bay, County Antrim, 113
Wicken Fen, Cambridgeshire, 36, 50, 265,
    286–7, 289, 291, 332
Wightwick Manor, near Wolverhampton,
    91–2, 162, 212
wildlife protection, 285–6
Wilkinson, Miss (Hamer's secretary),
    47–8
Wilks, John, 314–15, 318
Willes, Margaret, 263–4
Williams-Ellis, (Sir) Clough: supports
    CPRE, 61; on decline of country
    houses, 74; donates Snowdonia land
    to NT, 114; and postwar new towns,
    122; and NT's Welsh interests, 177–8;
    designs Cushendun cottages, 187;
    (ed.) *Britain and the Beast*, 82, 137;
    England and the Octopus, 61–2, 70
Wimbledon Common, 3
Wimpole Hall, Cambridgeshire, 276, 294
Winant, John Gilbert, 119
Windermere Lake, Cumbria, 29, 217
Wing, Buckinghamshire, 135
Winnifrith, Sir John, 241, 248–50, 255, 259
Wirksworth, Derbyshire, 105
Witley Common, Surrey, 288
Woburn Abbey, Bedfordshire, 173
Wodehouse, (Sir) P. G., 144
Wolfe, General James, 41
women: in senior positions, 309
Woodeson, Susan, 309
Woodward, Miss J. L., 35

Woolacombe Sands, North Devon, 38
Wordsworth, William, 9, 68
Working Men's College, 6
Works, Minister of: makes grant awards,
    165
Works, Office (*later* Ministry) of: Ancient
    Monuments Department, 74, 142
Worksop, Nottinghamshire: house and
    contents, 321
Worsley, Sir Marcus, 247, 301, 311–13,
    329
Wray, Cumbria, 9–10; Castle, 59
Wyatt, James, 183
Wycombe Peace Council, 301
Wyndham, John (*later* 1st Baron
    Egremont), 103, 146–7, 204, 226
Wynne-Finch, Colonel J. C., 182, 226–7

Yorke, Harriot, 7, 24, 95
Yorke, Philip, 275–6
Young National Trust Theatre, 267
Youth Hostels Association, 63, 235–7
Ysbyty estate, North Wales, 319, 321

Zetland, Lawrence John Lumley Dundas,
    2nd Marquess of: chairmanship of NT,
    74, 76–7, 96; qualities and personality,
    77; on Matheson, 81; and country
    house crisis, 82–3; speech in 1939
    National Trust Bill debate, 88; on
    timing of meetings, 97; and
    institutional occupancy of large
    houses, 99; Lees-Milne on human side
    of, 99; and Lloyd-Johnes donation,
    114; accepts Swanton Morley, 119; and
    postwar Labour government, 122–3;
    Princess Royal declines invitation to
    presidency, 123–4; resigns, 124;
    Crawford succeeds, 126–7; and NT in
    Northern Ireland, 182

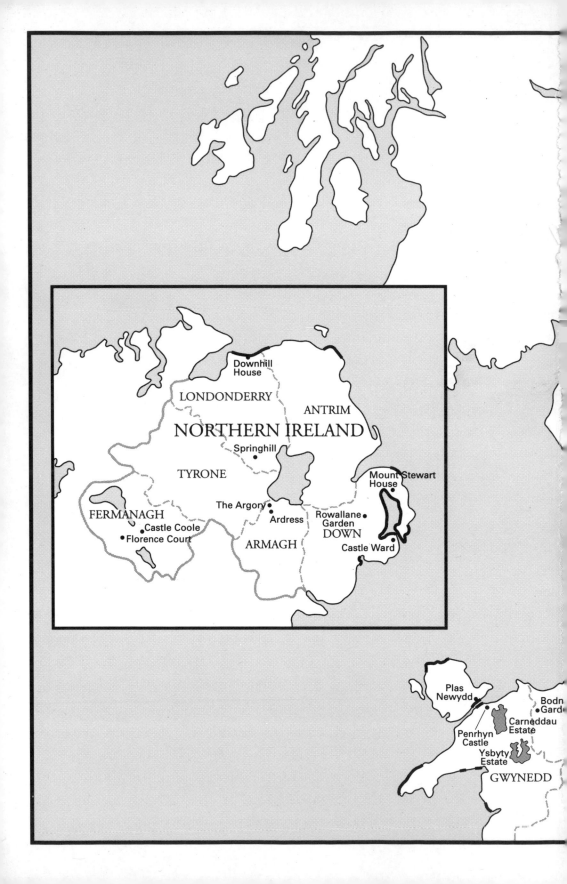

Downhill
House

LONDONDERRY

ANTRIM

NORTHERN IRELAND

Springhill

TYRONE

Mount Stewart
House

The Argony

FERMANAGH

Ardress

Rowallane
Garden

Castle Coole

DOWN

Florence Court

ARMAGH

Castle Ward

Plas
Newydd

Bodn
Gard

Carneddau
Estate

Penrhyn
Castle

Ysbyty
Estate

GWYNEDD